Climate Change and Agriculture

Climate Change and Agriculture

Dr. V. Radha Krishna Murthy Ph.D., PGDES

Former Senior Professor and Head, Department of Agronomy,
ANGR Agricultural University,
Agricultural College, Bapatla, A.P. India

Former Expert Team member in several Committees of CAgM of World Meteorological Organization, Geneva, Switzerland

Former International Agrometeorologist with Government of Nepal

Former Consultant with WB on Agrometeorology

Former Director, Roving Seminars on "Weather- Climate-Farmers"

Former INSAM Correspondent for Asia

BS Publications™

An Imprint of **BSP Books Pvt. Ltd.**

4-4-309/316, Giriraj Lane, Sultan Bazar,
Hyderabad-500 001, Telangana, India.

Climate Change and Agriculture ***by*** **V. Radha Krishna Murthy**

Published by:

BSP BS Publications™
An Imprint of **BSP Books Pvt. Ltd.**
4-4-309/316, Giriraj Lane, Sultan Bazar,
Hyderabad- 500 001, Telangana, India.
Phone: 040 - 23445688
e-mail: info@bspbooks.net
www.bspbooks.net

ISBN: 978-93-47311-48-2 (Paperback)

DISCLAIMER

The information provided in the book is not a substitute for professional advice and that readers assume all risks for their actions based on the content. The publication provides information on the subject matter, but the author is not offering professional services such as legal, investment, accounting advice, etc., related to climate science. It may also be noted that while efforts were made to ensure accuracy, no representations or warranties are made regarding completeness. To the maximum extent permitted by law, the author disclaims any and all liability in the event any information, commentary, analysis, opinions, advice and/or recommendations contained in this book prove to be inaccurate, incomplete or unreliable, or result in any investment or other losses.

The views and opinions expressed in this book are solely those of the author and do not represent any official stance or policy of any organization. The content is not intended to harm or defame any person, group, or institution. The information provided in this book is for general informational and educational purposes only.

While the author has made every effort to ensure the accuracy and completeness of the information presented at the time of publication, no guarantee is given regarding errors, inaccuracies, or omissions. The author assumes no responsibility for any liability, loss, risk, personal or otherwise, that is incurred consequently, directly or indirectly, of the use or application of any of the contents of this book.

This book is based on the best available science at the time of publication, using sources such as the Intergovernmental Panel on Climate Change (IPCC), WMO, FAO, UNEP, IMD, and peer-reviewed research. Climate science is constantly changing, and new discoveries may alter current understanding. The author does not guarantee the absolute accuracy, completeness, or reliability of the scientific information presented.

- Author

Dedication

My wife Smt. Padmavathi, son Srivenkatesh, daughter and son-in-law Smt. Anasuya Pooja and Chiranjeevi Krishna Kireeti, joined me in dedicating this book to The Almighty Lord Sri Tirumala Tirupati Sri Venkateswara Swamy Vaaru "The benevolent protector, The supreme preserver and savior, Lord of the seven Venkata Hills, The boon-giver of the Kali Yuga, Ocean of infinite auspicious qualities, Sea of compassion and mercy, Eternal protector of devotees, Lord of all worlds and realms, Possessor of a divine and auspicious form, Bearer of four arms symbolizing omnipotence, Bestower of all wealth and prosperity, and Ever-auspicious and eternally benevolent deity" for HIS copious blessings on us for our peace, prosperity, and happiness always and forever.

FOREWORD

Human activities are unequivocally causing global warming. Global warming has reached 1.1°C above pre-industrial levels, causing widespread impacts and risks to agriculture. Climate change impacts are intensifying on global agriculture and will continue to increase with further warming. The challenges ahead for agriculture globally are feeding an increasingly populous and hungrier world whilst managing increased risks from climate change on agriculture. In this direction, the book on "Climate Change and Agriculture" by Dr. V. Radha Krishna Murthy meets the knowledge to build capacity for climate action and enable design and implement readiness support effectively. Also, this book offers students, policy, and law makers a comprehensive lens into the dynamic relationship between agroecosystems and agricultural practices, blending scientific insight with real-world applications.

Designed to empower the next generation of thinkers and researchers, the first nine chapters explore the sources, trends, projections, and sector wise emissions of GHGs. The chapters from 10 to 15 covered strong agriculture resilience, while also highlighting innovative solutions from climate-smart farming systems to mitigation and adaptations in agriculture and livestock sectors. Global policy frameworks pertaining to climate change and agriculture covered in the last three chapters are appropriate.

His Concepts viz., "Murthy's Policy Concept on Climate Change (MPCCC) and "Murthy's Micromet Concept on Climate Change (MMCCC)" elaborately explained in the book are power packed with policy and classical science of micrometeorology. I strongly believe that these concepts help in addressing the negative impacts of climate change on global agriculture in general and developing and small island countries in particular. Faith moves mountains. Some Concepts move the Earth. Dr. Murthy did what I wished for from him.

I commend and congratulate the author of the book Dr. Murthy, a genius and born agrometeorologist, for his exceptional contribution to global education, research, and policy on climate change and agriculture.

Dr. M.V.K. Sivakumar

Former Director of the Climate Prediction and Adaptation Program and Former Acting Secretary of the IPCC, WMO, Geneva, Switzerland.

ACKNOWLEDGEMENTS

My wife typed the full manuscript of this book as well the way she did for my Doctorate thesis and earlier books. This book is as much hers as it is mine. My deepest gratitude and appreciation for her support in guiding this work and making it worthy. All auspiciousness in this work, my life, and career belongs to her.

My son Srivenkatesh has been my companion in the creation of knowledge. His actions always fulfill my sacred duties. His support is my life force and my prosperity. Through his devotion this book breathes. "Be glorious, my son grow in wisdom, uphold righteousness, speak truth, and may your path be ever auspicious".

The wisdom, patience, and unwavering support of my beloved daughter and son-in-law helped bring this book to life. Their presence is a blessing and helps are sacred offerings. May the fruits of this work bring light to them as their affection has brought light to me. "Be amazing my daughter and son in law- May your lives be filled with *ananda* (bliss), *shanti* (peace), and *sampatti* (prosperity). I honor your presence in my journey, as I honor the divine in you".

My ultimate tributes to my parents, in- laws, and all family members who instilled the best virtues, truth, compassion, and resilience in me. Also, I thank and bow to all my revered Gurus whom I always consider the embodiment of the Supreme Absolute for imparting me the knowledge to lead my life

My heartfelt thanks for the extraordinary guidance and support that Dr. M.V.K. Sivakumar garu provided throughout my career and writing foreword of this book. His insights, encouragement, and generosity have shaped not only my professional path but also my personal growth as an agrometeorologist. Working with him has been a privilege. His brilliance, humility, and dedication to advancing knowledge have inspired me more than words can convey. I thank him for sharing his wisdom so generously, and for being a role model of scientific excellence and integrity. I hope to carry forward the values he has instilled and contribute meaningfully to climate change and agriculture.

I gratefully acknowledge the insights and contributions of several anonymous authors whose work informed this study, even though specific references could not be located. This work has also benefited from ideas and perspectives shared in anonymous forums and publications where attribution was not possible. I am grateful for their insights, inspiration, and influence.

I am incredibly grateful to Shri. Nikhil Shah and Anil Shah of BS Publications for belief in my work and for bringing my book to life. Their support, guidance, and dedication throughout the publishing process have meant the world to me. I thank Mr. Naresh for helping me share my knowledge with the world.

-Author

SPECIAL ACKNOWLEDGEMENTS

The pioneer as an Hon'ble Minister, excellent leader, an ideal person, a soft voice like cold ice, marvellous State developer, top personality in technology skills, unbeatable icon, and people linguistic without any boundaries Late Sri. Mekapati Gowtham Reddy garu, is the son of Sri. Mekapati Rajamohan Reddy garu, a Legislator, Member of Parliament, a leading humanitarian and the doyen of farmers. Sri. Gowtham Reddy garu made his own mark in the fields of business, education, service, and politics of *Vasudaika Kutumbam.* Sir! Your support has been a cornerstone in advancing my earth-moving triad of concepts PCCC, MCCC and CCAC. Through this triad it is be possible to reduce the cost of cultivation of crops and livestock rearing by 50 per cent, increase in yields by 50 per cent, and another 10 per cent increase in quality of producc. The progress I achieved in this direction is to a large extent a direct reflection of your unwavering belief in my scientific capabilities. Your encouragement has provided me the momentum needed to bring my ideas in my concepts to life. I am deeply grateful for your insightful feedback and spirit throughout my journey. Your suggestions have been logistical and vital sources of inspiration and motivation. The dedication you've shown highlights your vision for innovation in my scientific field to help the food providing Gods, the farmers, irrespective of their affiliations. Thank you for championing my ideas when others saw only obstacles. Your commitment to pioneering new ground without boundaries is truly commendable. You strongly believed as much as I that my concepts definitely have lasting positive impact for profitable farming and food security.

-Author

PREFACE

"Satyameva Jayate" - "Truth alone triumphs". "All honors to you, Lord"

I prostrate before The Almighty Lord Sri. Tirumala Tirupati Sri Venkateswara Swamy Vaaru in highest obeisance with folded hands, offering my deepest gratitude for being the shakti behind this book.

HE destined me to study, spend all my life, learn, and teach my students about the most powerful forces of the Universe "Pancha Bhootas/Five Natural Forces" viz., Akasha/Aether (Solar radiation), Vayu (Air), Agni (Temperature), Apah/Jal (Rainfall), and Prithvi (Cultivated soil) and agriculture. I offer my deepest gratitude to these forces as well whose presence and inspiration have guided this work.

Climate Change is happening. It is negatively impacting agriculture and livestock. The cost of cultivation of crops and animal rearing is alarmingly increasing beyond the control and affordability of the farmers and returns are meagre, forcing the farmers to take extreme steps. Scientific works and studies project that climate resilient crop production, i.e., by monitoring daily weather conditions; following weather and climate information; services and products; and adopting micrometeorological technics during the crop growing season it is possible to reduce cost of cultivation of crops and animal rearing by 50 per cent; increase in yields by 50 per cent; and 10 per cent in quality improvement of crops and livestock. By linking PCCC with MCCC, and CCAC discussed in the book, it is very much possible to achieve these benefits and farmers income doubles under climate change scenarios.

Over the years I synthesized the information, analyzed the findings, and documented my experiences related to climate change on agriculture and climate action and taught my students. The same material is embedded in the form of the book "Climate Change and Agriculture". May this humble offering by me serve the greater good and reflect the light of truth.

"Satyameva Jayate" - "Truth alone triumphs". "All honors to you, Lord"

- Author

CONTENTS

Chapter 2

Greenhouse Gases (GHGs) and Climate Change

Chapter 3

Climate Change and Solar Radiation

Chapter 4

Climate Change and Wind

Chapter 5

Climate Change and Temperature

Chapter 7

Climate Change and Sea Level Rising and Extreme Events

Chapter 8

Climate Change and Photosynthesis and Cultivated Soils

Chapter 9

General options for Climate Change, Climate Neutral, and Net Zero Emissions

Chapter 10

Climate Change and Global and Indian Agriculture

Chapter 11

Climate Change and Major Cereal and Pulse Crops

Chapter 12

Climate Change and Major Oilseeds and Commercial Crops, and Pests and Diseases

Chapter 14

Climate Change and Livestock and Humans

Chapter 15

Climate Change, Food Production and Food Security, and Role of Agrometeorology

Chapter 16

Global Conventions, Treaties, Summits, and Schools of Thought on Climate Change

Chapter 17

Agrometeorological Concepts for Climate Change and Agriculture

Chapter 18

Project on Establishment of Climate Change and Agriculture Center (CCAC)

LIST OF TABLES

LIST OF FIGURES

Chapter-5

Chapter-6

Chapter 15

Chapter 17

Chapter 1

Climate Change and Variability

"The Vedas describe the atmosphere as a sacred and dynamic realm, a crucial intermediary between Earth and the celestial heavens. It is a domain permeated by divine energies, influencing both nature and human existence. Through hymns and rituals, the Vedic sages acknowledged its profound significance, viewing it as essential for life, well-being, and spiritual endeavors".

1.1 Introduction to Climate Change

- Climate change is the widely accepted single largest pressing issue facing society
- Climate change has been and continues to be the principal source of fluctuations in agriculture, food production, and food security
- Throughout history, extreme events such as droughts, floods, cyclones, heat and cold waves, and various forms of extreme weather and climate caused enormous damage to the agricultural systems
- In conjunction with other physical, social, political, and economic factors, climate change contributes vulnerability to hunger, famine, dislocation, and economic losses
- Hence, it is imperative that climate change aspects are well understood to formulate more sustainable policies and strategies to promote agriculture and food production, thereby food security

- Increased emissions of Green House Gases (GHGs) through anthropogenic activities are causing climate change
- ***The Earth System***
    - The Earth system comprises interconnected components, each influencing the others. This includes the biosphere (living organisms, including crops), the atmosphere (air and gases) the pedosphere (soil) and climate change impacts these elements that interact among themselves **(Fig.1.1)**

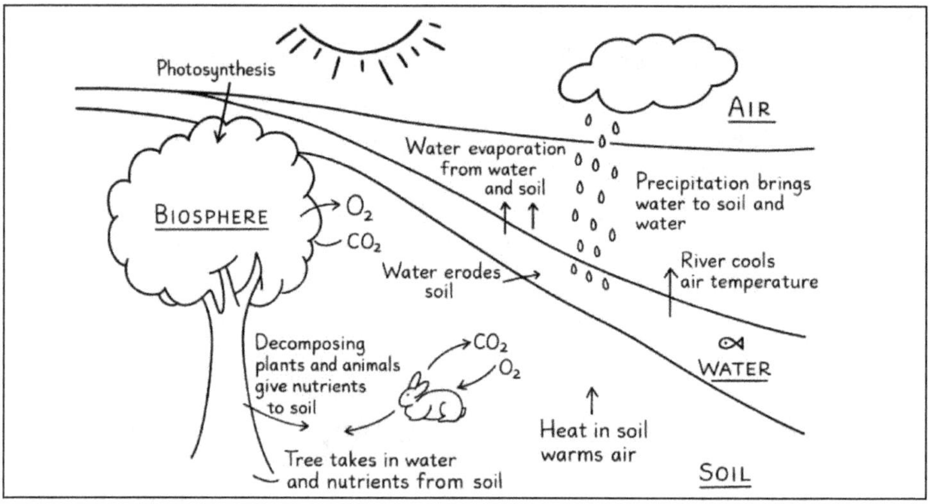

Fig. 1.1 The earth system.

Source: Unknown. Class lecture notes collection.

- Biosphere and atmosphere: Trees/ crop plants (biosphere) absorb carbon dioxide from the atmosphere during photosynthesis, converting it into biomass and oxygen. Animals respire oxygen and release carbon dioxide. This exchange helps regulate climate and support life. Human activities, like burning fossil fuels and land-use change, disrupt this balance by increasing greenhouse gases in the atmosphere, leading to climate change
- Soil and atmosphere: Soil and atmosphere interact through various biogeochemical cycles. Soil absorbs solar energy and transfers it to the atmosphere, influencing climate patterns. Soil also plays a vital role in the carbon cycle,

storing a significant amount of carbon. Climate change, particularly increased temperatures, can accelerate carbon loss from soils into the atmosphere, creating a feedback loop that exacerbates warming

- Trees/crops, biosphere, and soil: Trees, crops, as part of the biosphere, depend on healthy soil for essential ecosystem services like nutrient cycling and water retention. However, agricultural practices significantly influence soil health and the weather health

- ***Climate of the Earth***
  - The climate is a complex, interactive system consisting of (a) atmosphere, (b) land surface (living things, snow, and ice), and (c) oceans and other bodies of water
  - The atmospheric component of the climate system characterizes climate
  - Climate is described in terms of the mean and variability of temperature, rainfall/precipitation, and wind over a period ranging from months to millions of years (the classical period is 30 years)
  - The climate system evolves under the influence of its own internal dynamics and due to changes in external factors that affect climate (forcings)
  - Solar radiation powers the climate system
  - Climate is renewable source of energy
  - Weather and climate information is non- monetary input in agriculture

- Definition: Climate change is a change in the state of the climate that can be identified (using statistical tests) by changes in the mean and/or the variability of its properties and that persists for an extended period, typically decades or longer
  - *The climate change is due to*
    - Natural internal processes
    - External forcings
    - Persistent anthropogenic changes
    - Anyone or combination of the above processes cause changes in the composition of atmosphere and land use

- Relationship between Climate Change and Weather
    - o Climate change and weather are intertwined (average weather is climate)
    - o Observations show that there have been changes in weather and it is the statistics of changes in weather over time that identify climate change
    - o The chaotic nature of weather makes it unpredictable beyond a few days
    - o Predicting climate change due to changes in atmospheric composition and other factors is much more meaningful issue than weather prediction which is a very different issue
    - o While many factors continue to influence climate, scientists have determined that human activities have become a dominant force, and are responsible for most of the warming observed over the past fifty years
    - o Human-caused climate change has resulted primarily not only from increases in the amounts of greenhouse gases in the atmosphere but also from changes in small particles (aerosols) as well as from changes in land use
    - o As climate changes, certain types of weather events are affected. Example: As earth's average temperature increased, the weather phenomena like heat waves, heavy rainfall/ precipitation, etc., have become more frequent and intense, while others like extreme cold events have become less frequent and intense

1.2 Significance of Climate Change for Agriculture

- Climate change presents a complex and significant challenge to agriculture. Its influence ranges from crop yields and livestock health to weather health, water resources, soil health, and pest management. The effects are multi-faceted and vary regionally, making climate change a critical concern for farmers, policymakers, and consumers alike
- Changes in crop yields: Rising temperatures and alterations in rainfall patterns can significantly impact crop growth and

development, potentially leading to changes in yields for many major crops like maize, rice, wheat, soybean, cassava, and oats. Some regions may experience slight yield gains for crops like wheat, especially in higher latitudes with cooler temperatures that could become more suitable for agriculture. However, even in areas with potential benefits, those gains may be offset by the increased frequency of extreme events like heat waves and droughts

- Heat stress: Dairy cows, for example, are particularly vulnerable to heat stress, which can negatively affect their appetite, milk production, and overall health
- Pests and diseases: Warmer temperatures and changing precipitation patterns can create conditions favorable for the expansion and increased virulence of agricultural pests and diseases, necessitating greater pesticide use and potentially reducing yields
- Water resources: Climate change is disrupting the water cycle, leading to shifts in precipitation patterns and increased evaporation rates. This can result in increased water scarcity in some areas, potentially limiting the availability of water for irrigation. Conversely, other regions may experience increased runoff and flooding due to heavier rainfall events, which can erode soil and affect crops
- Soil health: Intensified rainfall and increased storm frequency can contribute to soil erosion, potentially depleting vital nutrients and degrading soil quality. Rising temperatures can also affect the ability of soils to store carbon exacerbating climate change through a feedback loop
- Weather health: Injection of GHGs into the atmosphere adversely affect weather health thereby negatively impacting "Soil- Plant- Weather Continuum"
- Economic strain on farmers: Declining yields, increased costs for managing pests and diseases, and investment in mitigation and adaptation measures can strain farmers financially, potentially leading to reduced profits

- Impacts on food security: Reduced agricultural productivity and increased food prices can have severe consequences for food security particularly in vulnerable regions already grappling with poverty and hunger
- The significance of climate change for agriculture cannot be overstated. It is a fundamental force reshaping agricultural landscapes and posing substantial threats to food security, the livelihoods of farmers, and the delicate balance of ecosystems. Understanding these impacts and actively pursuing mitigation and adaptation strategies are essential for building a resilient agricultural sector capable of providing a stable food supply in a changing world

1.3 Basics on Computing and Calculation of Climate Change

- The Intergovernmental Panel on Climate Change (IPCC) provides comprehensive assessments of scientific, technical, and socio-economic information related to climate change. While they do not conduct their own research, they synthesize and assess published scientific literature to inform policymakers and the public
- The IPCC's methods in computing climate change
- Data and observations
    - Reliance on observational datasets: The IPCC assesses global temperature change by analyzing readings from thermometers worldwide, incorporating statistical techniques to account for sparse data areas like the poles. They utilize long-term series of temperature measurements from land-based stations, ships, buoys, and satellites
    - Paleoclimate data: Data from natural archives like ice cores, tree rings, and sediment layers provide information about past climates and help understand the natural variability and long-term trends

- Climate models and scenarios
  - Global climate models (GCMs): These complex computer models simulate Earth's climate system, incorporating atmospheric, oceanic, land, and cryospheric processes. They are crucial for projecting future climate changes under different emission scenarios
  - Climate forcing (radiative forcing): This metric quantifies the impact of various factors (like greenhouse gases, aerosols, and land use changes) on Earth's energy balance and thus its climate system
  - Emission scenarios: The scenarios developed with socioeconomic factors in mind, provide different projections of future greenhouse gas emissions based on assumptions about population growth, technology, and policy choices. They help evaluate the potential consequences of different actions or inaction
  - Climate sensitivity: This refers to the global surface temperature warming at equilibrium in response to a doubling of atmospheric CO_2 concentrations. IPCC reports provide a range for this sensitivity, reflecting the uncertainties in modeling complex feedback mechanisms

- Assessment and reporting
  - Expert assessment: Thousands of volunteer scientists from around the world contribute to the IPCC's assessment reports, evaluating thousands of scientific papers each year
  - Multi-stage review process: IPCC reports undergo a rigorous, multi-stage review process by experts and governments to ensure objectivity, completeness, and a diverse range of views and expertise
  - Key findings and uncertainty language: The IPCC identifies the strength of scientific agreement in different areas and uses calibrated uncertainty language to communicate the degree of certainty in its findings
  - The IPCC's work is crucial for understanding the science of climate change, informing policy decisions, and guiding global efforts to address this critical challenge

- Calculation of Climate Change
  - Based on the information available through the IPCC (Intergovernmental Panel on Climate Change) and related sources, some related equations are used for calculating temperature and other climate change factors
- Earth's energy budget
  - The Earth's temperature is fundamentally linked to its energy budget, the balance between incoming solar radiation/shortwave radiation and outgoing infrared/longwave radiation
  - Changes in this balance, often caused by the release of greenhouse gases like carbon dioxide and aerosols, lead to changes in global temperature
- Radiative forcing (ΔF)
  - This refers to the change in net radiation at the top of the atmosphere due to a particular factor, like an increase in greenhouse gas concentrations
  - A simplified logarithmic equation, commonly cited in relation to IPCC findings, relates a change in CO_2 concentration to radiative forcing $(\Delta F) = 5.35 \ln (C/C_0)$
  - ΔF: Increase in radiative forcing in Watts per square meter
  - ln: Natural logarithm
  - C: Current concentration of atmospheric CO_2 in parts per million (ppm)
  - C_0: Pre-industrial CO_2 concentration (280 ppm in 1750)
  - 5.35: A scalar value derived from radiative transfer calculations
- Temperature increase (ΔT)
  - The increase in radiative forcing can be translated into a global temperature increase
  - A simplified equation (though acknowledging the complexities of climate feedback mechanisms) can relate radiative forcing to temperature change $= \lambda * \Delta F$
  - ΔT: Temperature increase in degrees Celsius (or Kelvin)

- λ: Climate sensitivity parameter (often denoted as lambda), indicating the global temperature response to a change in radiative forcing
- ΔF: Increase in radiative forcing (as calculated above)

- Other variables and complexity
  - Comprehensive climate models used by the IPCC incorporate numerous equations to represent complex interactions within the Earth system, including
    - Navier-Stokes equations for fluid motion (winds, ocean currents)
    - Thermodynamic equations describing heat and moisture transfer
    - Equations for other climate variables such as precipitation, sea level rise, and ice loss
  - These complex models are crucial for capturing the regional and global details of climate change, including feedback (processes that amplify or dampen initial warming) that affect the final temperature response
  - The simplified equations presented above are useful for understanding the fundamental relationships but represent a highly simplified view of the complex calculations involved in detailed climate modeling

1.4 Climate Change, Global Processes, and Effects

- Climate change is the defining issue of the present era. Hardly a day passes without a newspaper, a broadcast or a policy maker, a lawmaker, etc., making at least one statement or reference to the threats climate change poses and the urgency of taking action immediately to limit the effects and in the long term to adapt to the changes that are sure to come. Highly perceptible changes in climate can occur locally even over a few centuries. Anthropogenic activities have brought, within the realm of reality, the possibility of perceptible climate changes even within decades. The climate change, global processes, and effects are as follows

- **Human activities**
  - Land Use Change
    - *Urbanization*
    - *Deforestation*
    - *Land conversion to agriculture*
    - *Increase in impermeable surface*
  - Fossil fuel burning
    - *Industry*
      - Chemicals
      - Cement
    - *Energy production*
      - Heating
      - Electricity
      - Power
    - *Transport*
      - Plane traffic
      - Shipping
      - Road
      - Rail
    - *Agriculture*
      - Fertilizers
      - Chemicals
- **Climate Change Processes**
  - Enhanced Greenhouse effect
    - *Carbon cycle disturbances*
    - CO_2
    - CH_4
    - N_2O
  - Global Warming (average temperature rise)
    - *Ice Caps melting (sea level rise)*
    - *Precipitation changes*
    - *Cloud cover changes*
    - *Ocean circulation upheaval*

    - ❖ Salinity
    - ❖ Water temperature
  - *Gulf stream modification*
  - *Monsoon disturbances*

- **Disasters**
  - o *Environmental refugees*
    - *Droughts*
    - *Cyclones*
    - *Floods*
    - *Tsunami*
    - *Wildfire*
  - o *Biodiversity losses*
    - *Coastal wetlands disappearing*
    - *Coral bleaching*
  - o Casualties
  - o Economic losses
  - o Diseases spread
    - *Infectious diseases* (*vector change*)
    - *Diarrhea*
    - *Cardio-respiratory diseases*
  - o Subsistence farming and fishing at stake
  - o Malnutrition
  - o Traditional lifestyles
  - o Coastal wetlands disappearing

1.5 The Greenhouse Effect

- The glass walls in a greenhouse reduce airflow and increase the temperature of the air by trapping large quantities of energy inside. Analogously, but through a different physical process, the earth's greenhouse effect warms the surface of the planet
- Global budget of incoming solar radiation and "Natural greenhouse effect"

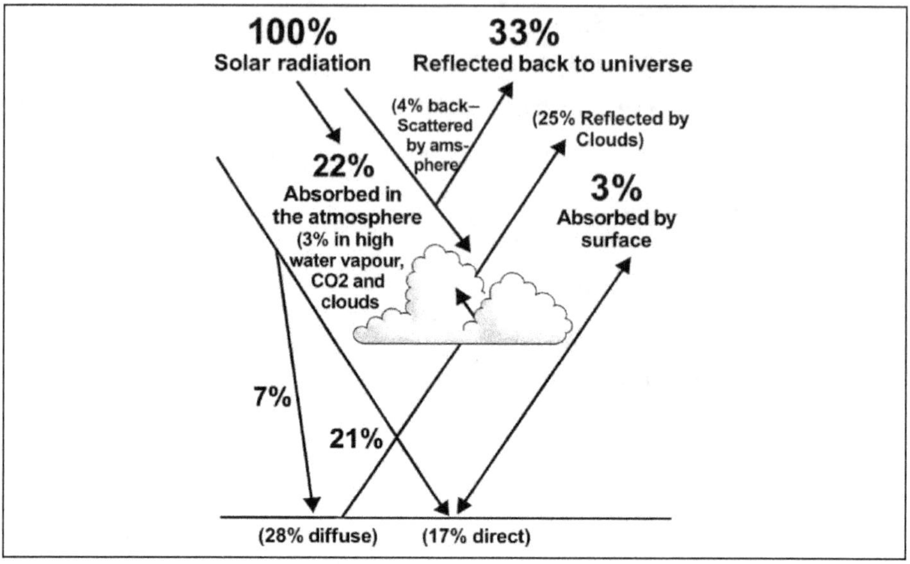

Fig. 1.2 Global budget of incoming solar radiation.

Source: Unknown. Class lecture notes collection.

- Global budget of incoming solar radiation refers to "The balance between incoming solar radiation and outgoing terrestrial radiation" **(Fig:1.2).** It also refers to "The equilibrium between incoming solar radiation and outgoing terrestrial radiation ensuring a relatively stable global temperature"
- Incoming solar radiation is the primary energy source for the Earth's climate system, in the form of shortwave radiation (UV, visible light, and a portion of infrared)
- Roughly 33% of this incoming radiation is reflected to space by clouds, the atmosphere, and Earth's surface (snow, ice). This reflectivity is called "Albedo"
- The remaining 67% is absorbed
  - 22% absorbed by the atmosphere i.e., 19 per cent by water vapor, carbon dioxide, clouds and 3 per cent in high atmosphere
  - 45% absorbed by the Earth's surface (land and oceans)

- The absorbed energy is re-radiated as longwave radiation (infrared energy)

- Greenhouse gases (carbon dioxide and water vapor) absorb much of this re-radiated longwave radiation, trapping heat and warming the Earth's atmosphere and surface
- This is known as "Natural greenhouse effect" without which the Earth would be significantly colder, making it uninhabitable
- Therefore, the "Earth's energy balance" which ensures "Natural greenhouse effect "is crucial for maintaining a habitable temperature

1.5.1 Greenhouse Gases and "Enhanced Greenhouse Effect"

- Human activities are increasing the concentration of greenhouse gases leading to "Enhanced greenhouse effect" **(Fig:1.3)**
- Enhanced Greenhouse Effect: The two most abundant gases in the atmosphere, nitrogen (comprising 78 per cent of the dry atmosphere) and oxygen (comprising 21 per cent) exert almost no greenhouse effect. However, human activities, such as land use change, burning fossil fuels, etc., release large amounts of greenhouse gases into the atmosphere intensifying the greenhouse effect, which is known as "Enhanced greenhouse effect"

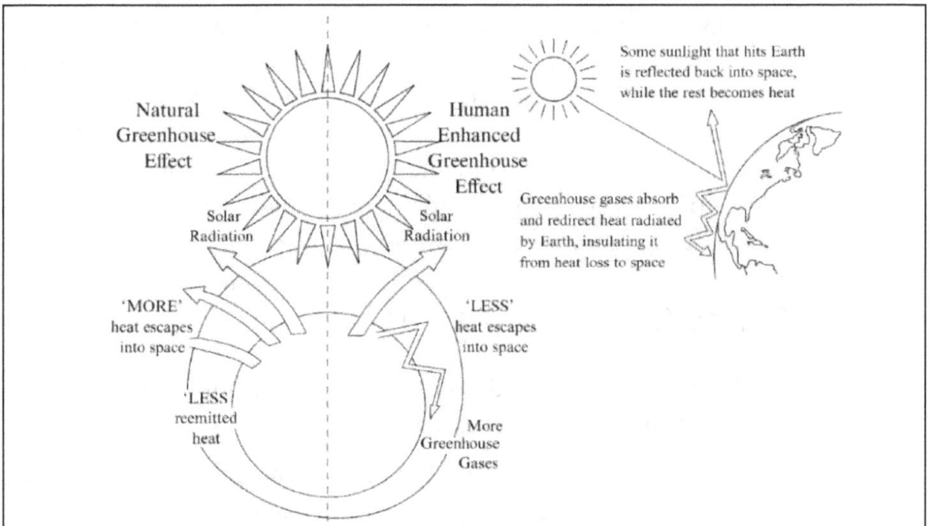

Fig. 1.3 Greenhouse effect.

Source: Unknown. Class lecture notes collection.

- In the natural greenhouse effect more heat escapes into the space and less re-emitted heat is present in the atmosphere
- In the human enhanced greenhouse effect less heat escapes into the space and more re emitted heat enters into the atmosphere
- The “Enhanced greenhouse effect” leads to an imbalance in the Earth's energy budget causing “Global warming” which results in “Climate change”
- In turn, the “Climate change” results in “Changes in weather patterns”
- In essence, the “Natural greenhouse effect” is a natural phenomenon that keeps Earth warm enough for life. However, human activities are amplifying this effect, resulting in “Enhanced greenhouse effect” leading to a shift in the planet's energy balance, global warming, and climate change which have potentially harmful consequences.

1.6 Climate Change and Global Warming

- “Global warming” is a major contributor to “Climate change” which encompasses a broader range of changes in weather patterns, sea levels, and other climate-related phenomena. Global warming refers to the increase in global surface temperature relative to a baseline reference period, averaging over a period sufficient to remove interannual variations (20 or 30 years). A common choice for the baseline is 1850–1900 (the earliest period of reliable observations with sufficient geographic coverage) with more modern baselines used depending upon the application

1.7 Climate Change and Air Pollution

- Climate change and air pollution, while distinct, are interconnected to some extent
- Air pollution is primarily caused by the release of harmful substances into the atmosphere which can contribute to climate change locally by releasing greenhouse gases like carbon dioxide and methane

- Conversely, climate change can exacerbate air pollution
- Air pollution represents a significant global environmental challenge with profound implications for agricultural productivity and sustainability
- Increased temperatures and wind speeds decrease pollutants
- Particulate matter reduces photosynthetic efficiency through physical blockage for gases exchange and stomatal conductance
- Sulphur dioxide and nitrogen oxides cause direct foliar damage and soil acidification mechanisms
- There is a critical need for integrated air quality monitoring in agroecosystems

1.8 Global Sector Wise Greenhouse Gases Contributions to Climate Change

- The combustion of coal, oil, and natural gas for electricity generation, transportation, and industrial processes emit substantial quantities of carbon dioxide, constituting the foremost Green House Gases (GHGs) and account for large portion of global GHG emissions. Agriculture and land use changes including rice cultivation, livestock rearing, and fertilizer application emit methane and nitrous oxide add further and account for second largest contribution to global emissions. Deforestation and land use alterations, which include the clearance of forests for agriculture, urban expansion, and other human endeavors liberate stored carbon into the air and release GHGs to Earth's Climate also substantially account for global emissions
- Energy supply: 34%
  - This sector is the largest contributor to global greenhouse gas emissions. It includes the burning of fossil fuels (coal, natural gas, and oil) for electricity and heat production. In 2023, the global power industry alone produced 15.1 gigatons of carbon dioxide equivalent ($GtCO_2e$) accounting for 28% of global emissions

- Industry: 24%
    - This includes emissions from industrial processes like those involved in producing chemicals and cement and emissions from burning fossil fuels on-site at facilities for energy. Emissions from industrial processes have seen the fastest growth since 1990 increasing by 225%
- Agriculture, Forestry, and Other Land Use (AFOLU): 22%
    - Emissions from this sector mainly stem from agricultural activities (cultivation of crops and livestock) and deforestation. AFOLU emissions are subject to higher uncertainty and annual variability compared to other sectors, making their long-term trend harder to determine
- Transportation: 15%
    - This sector is primarily driven by burning fossil fuels for various forms of transport. Road travel makes up most of these emissions
- Other sectors including buildings: 5%
    - Emissions from this sector come from onsite energy generation and fuel burning for heating and cooking. When electricity use is included this sector's contribution increases significantly

1.9 Global Transport Sector Contributions to Climate Change (2019 Data)

- The transportation sector is a major contributor to climate change, and its emissions by mode of transport are
- Road Transport
    - Dominant Source: Road vehicles are by far the largest source of transport emissions, accounting for approximately 70% of global transport-related CO_2 emissions

- Shipping (maritime transport)
    - Significant Contributor: International shipping is the second largest source of transport emissions, contributing around 11% of global transport-related CO_2 emissions
- Aviation (air transport)
    - Growing Emissions: Aviation accounts for roughly 12% of global transport-related CO_2 emissions and is experiencing rapid growth
- Rail and other Transport
    - Lowest emissions: Rail transport is a relatively minor contributor to global transport emissions, accounting for around 1% of the sector's total. Other Transport: (pipelines, etc.) approximately 6%
- In Summary
    - Road: 70%; Shipping: 11%; Aviation: 12%; Rail: 1%; Other Transport: (pipelines, etc.) approximately 6%

1.10 Global Agriculture Sector Contributions to Climate Change

- Globally, the food system is estimated to contribute a significant portion of greenhouse gas emissions, with figures ranging from 25% to 37% and is expected to increase further due to population growth and changing dietary patterns. The main contributors to greenhouse gas emissions from agriculture are
- Livestock and Fisheries
    - These activities are responsible for roughly 31% of emissions from the food system, primarily due to methane from livestock (especially ruminants like cattle), manure management, and fuel consumption from fishing vessels
- Crop production
    - Emissions from crop production for human consumption and animal feed (including fertilizer use and other on-farm processes) account for about 27% of food system emissions

- Land use change
    - This category, including deforestation for agricultural expansion and emissions from cultivated organic soils constitutes approximately 24% of the food system's emissions
- Supply chains
    - Post-farm activities like processing, transportation, and retail account for roughly 18% of emissions

1.11 Global Agricultural Chemicals Sector contributions to Climate Change

- Agricultural chemicals, particularly fertilizers and pesticides, significantly contribute to greenhouse gas (GHG) emissions throughout their entire life cycle from manufacturing to application and eventual environmental fate
- Fertilizers
    - Nitrous Oxide (N_2O)
        - Direct Emissions: Application of synthetic nitrogen fertilizers on agricultural land leads to the release of N_2O, a powerful GHG, into the atmosphere
        - Indirect Emissions: Excess nitrogen from fertilizer runoff enters waterways, leading to eutrophication and further N_2O emissions from aquatic systems
- Methane (CH_4) and Carbon Dioxide (CO_2)
    - Manufacturing: The industrial process for producing synthetic nitrogen fertilizers, like the Haber-Bosch process, uses fossil fuels (often natural gas, which is mostly methane) and releases both CH_4 and CO_2
    - Eutrophication: Over-fertilization contribute to eutrophication in bodies of water, which also release CH_4 and CO_2 as algae and other aquatic organisms decompose
- Pesticides
    - Manufacturing and transport: Nearly all synthetic pesticides come from fossil fuels. The manufacturing

process is energy-intensive, and their transport and packaging also contribute to GHG emissions. Producing one kilogram of pesticide requires approximately ten times more energy than one kilogram of nitrogen fertilizer

- Volatilization: Some pesticides, such as sulfuryl fluoride, are potent GHGs. Emitting one ton of sulfuryl fluoride equals the emission of nearly 5,000 tons of CO_2
- Ground-level Ozone: Many pesticides contribute to ground-level ozone formation, a harmful GHG and air pollutant
- Impact on soil health: Pesticides disrupt natural soil processes. This affects the soil's ability to store carbon and influences microbial communities that regulate GHGs

1.12 Global Economic Losses due to Climate Change

- Studies indicate that the estimated world economic losses due to climate change could be between 127 and 616 trillion dollars by 2100 with current commitments. Failure to implement current commitments raises economic losses to 150-792 trillion dollars by 2100

1.13 Country wise GHG emissions in GtCO_2e (Gt CO_2 e stands for Gigatons of Carbon Dioxide Equivalent)

- China-15.9; USA- 6.0; India-4.1; European Union-3.2; Russia-2.7; Brazil-1.3; Indonesia-1.2; Japan-1.0; Iran-1.0; Saudi Arabia-0.8; Canada-0.7; Mexico-0.7; South Korea-0.7; Turkiye-0.6

1.14 Global Regions most Vulnerable to Climate Change

- Globally, four (4) regions have been identified as the most vulnerable to climate change. These are as follows

- The Arctic
  - The causes of enhanced vulnerability in the Arctic are due to large regional warming and associated ice melt
  - The habitat of plant and animal species living along the ice margins such as polar bears will alter as the ice melts. The area of ice is already decreasing since 1979, and the Arctic has been losing summer ice at about 9 per cent per decade, and this trend is expected to continue
  - Traditional ways of life of indigenous inhabitants are likely to be affected (those of the Inuit which are based on hunting)
  - Ice melting associated with global warming leads to improved transport links and opportunities to open the Arctic to commercial activities, especially the exploitation of its mineral and oil wealth
  - However, these opportunities must be set against the threat to ecosystems and indigenous livelihoods and the risk of extinction of some iconic species
- Africa
  - Much of Africa is already vulnerable to drought due to high natural variability in the climate and high demand on natural resources from a population which is heavily dependent on agriculture for its livelihood
  - Thus, Africa is vulnerable to any increase in the frequency of occurrence and /or severity of drought due to global warming
  - The IPCC Fourth assessment considered it likely that the area affected by drought would increase during 21st century
  - These risks are compounded by structural problems, the quality of governance, and the health and education levels of the population limiting the capacity to adapt to climate change
- Asian and African mega deltas
  - The Asian and African (principally the Nile) mega deltas are areas of very high population density concentrated in

low- lying areas prone to flooding and exposed to storm surge and tropical storms
  - These areas are vulnerable to rising sea levels related to global warming and the severity of tropical storms and storm surge
  - By 2100 sea level is expected to rise substantially
  - The IPCC Fourth Assessment considered it likely that intense tropical cyclone activity will increase in the future

- Small islands
  - Small islands are also vulnerable to rising sea levels and increasing storm severity
  - In future higher mean sea levels may be sufficient to inundate some small islands
  - In the Caribbean and Pacific islands more than 50 per cent of the population live within 1.5 km of shore
  - Unlike deltas and other coastal areas small islands have no hinterland to move to in case of coastal land loss
  - As the sea level rises, it is not only coastal inundation that poses a threat to small islands but also contamination of freshwater supplies by salinization of the ground water may threaten the settlements
  - Tourism is a major contributor to the economies of many small islands. Beach erosion, degradation of coral reefs, and bleaching make small islands a less attractive destination for tourists

1.15 Evidence for Climate Change

- The following observed changes in the climate system since the late 19th Century reveal an unequivocally warming world
- Air: Increases in atmospheric circulation, air temperature, water vapor, and surface temperature
- Land: Increases in species ranges shift, growing season length, precipitation, and greenness
- Ocean: Increases in sea level, sea surface temperature, species ranges shift, ocean heat content, and ocean acidification

- Ice: Decreases in permafrost extent, ice sheets, Arctic Sea ice, snow cover, and glaciers

1.16 Climate Variability

- Definition: Deviations of climate variables from a mean state, including the occurrence of extremes, across all spatial and temporal scales beyond individual weather events
- The climate variability is due to
    - Natural internal processes within the climate system (internal variability)
    - Variations in natural or anthropogenic external forcings (external variability)
- Sources of Climate Variability
    - The sources of climate variability are represented on “Time and spatial scales”. Examples: A few days, Intra season, Inter season, A very few years, etc.
    - The day-to-day weather variations which are primarily controlled by internal atmospheric dynamics
    - Climate variations in intra- season, inter -seasonal, a very few years etc., are controlled by boundary conditions at the interface among atmosphere, ocean, and land surface
- Types of climate variability
    - *Short term variability:* In this type, the climate variability is confined to limited time and space. Examples: Cyclones, hurricanes, typhoons, droughts, floods, etc.
    - *Long term variability*: In this type, the variability is extended to relatively longer time and larger areas. Examples: Inter and intra annual variability in rainfall which is the key climatic element that determines the success of agriculture in a region; Variability in temperature during the whole crop season which impacts photosynthesis thereby final crop yield
- Approaches in climate impact assessment
    - Different approaches exist to assess climate impact. The specific terms "Impact Approach" and "Interaction

Approach" are always explicitly used to resonate with common ways to categorize such assessments **(Fig:1.4)**

- o Impact approach (top-down)

- *Focus*: This approach typically starts with projections of climate change (temperature and precipitation changes from climate models) and then assesses the direct consequences or impacts on a specific sector, ecosystem, or region
- *Methodology*: Involves modeling how changes in climatic variables, such as temperature and precipitation, would affect various systems (crop yields, water resources, sea level rise) without necessarily considering the socio-economic context or human responses

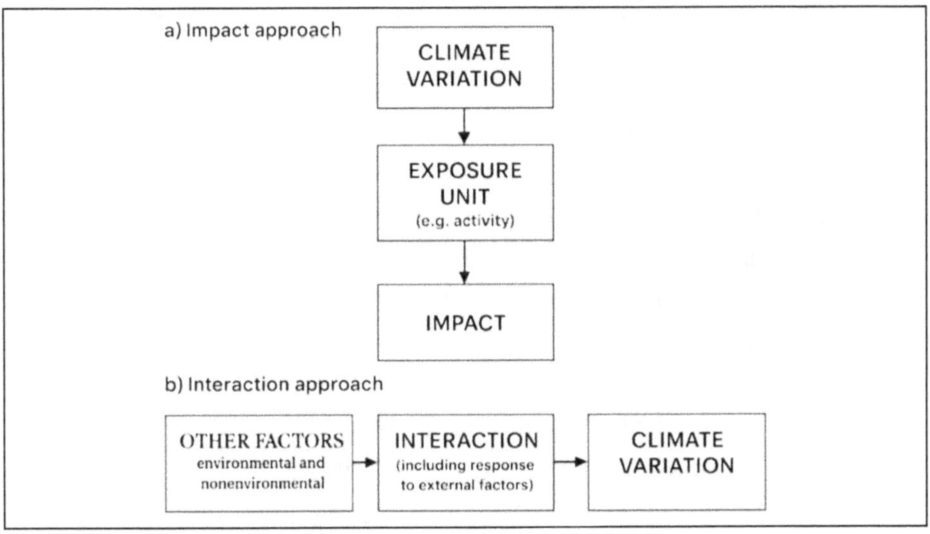

Fig. 1.4 Approaches in climate impact assessment.

Source: Parry and Carter,1989.

- *Strengths*: Allows systematic analysis of potential consequences under different climate scenarios. Can highlight areas of particular vulnerability to climate change
- *Weaknesses*: Can be seen as "Predict-then-act" or "Top-down," often making it difficult to fully integrate with local socio-economic contexts and adaptation strategies. May overlook the

dynamics and resilience of socio-economic systems and their capacity to adapt to changing conditions

- Interaction approach (bottom-up/co-evolutionary)
    - *Focus*: Emphasizes the interplay between climate variation and other factors viz., environmental and non-environmental (socio-economic factors), recognizing that these are not independent but constantly influencing each other. This approach prioritizes understanding system vulnerabilities and adaptive capacities in the context of both climatic and non-climatic stresses
    - *Methodology*: Often starts with identifying current climate risks and system sensitivities and explores adaptation options through participatory and scenario-based methods, working closely with local stakeholders. It considers how societal changes (independent of climate change) influence vulnerability and resilience
    - *Strengths*: More directly relevant for adaptation planning as it focuses on vulnerabilities and local needs. Fosters learning and engagement among stakeholders, promoting a more holistic understanding of climate risks and opportunities for action
    - *Weaknesses*: Can be more complex to implement and may require more intensive data collection and stakeholder engagement. Dealing with the uncertainties in both climate and socio-economic futures can be challenging
    - While the Impact Approach focuses primarily on the direct consequences of climate change, the Interaction Approach recognizes that climate change impacts are mediated and shaped by ongoing societal and environmental processes, necessitating a more integrated and adaptive approach to assessment of pests, diseases, and final crop yield.

1.17 Differences between Climate Variability and Climate Change

Table 1.1 Differences between climate variability and climate change.

S.No.	Climate variability	Climate change
1.	Observed short term changes/ trends of day to day, inter and intra seasons, year to year differences, etc. in values of specific climate variables within an averaging period, typically 30 years. Example: Variability in the annual rainfall of India from 1951 to 1990	Long term changes/trends between averaging periods either in the mean values of climate variables or in their variability. Example: Change in the global temperature for a period of 1915-1995 over a base period of 1750-1900
2.	Climate variability is a consequence of global climate change at local and regional levels. A notable consequence of climate variability is "Drought", which sharply increase crop failures, hunger and famine	Global climate change leads to regional climate variabilities. A notable consequence of climate change is "Sea level raising" which causes inundation of coastal areas and dislocation of inhabitants
3.	Ups and downs in day to day, season to season or year to year	Gradual and persistent shift in average climatic conditions
4.	Short term trends associated with natural variations which occur within shorter time frames such as days, weeks, months, seasons, and few years	Long term trends associated with natural forcings, external forces, and human activities or a combination of one or more
5.	The range and frequency of shocks that society can absorb or to which it adjusts	The range and frequency of shocks that society cannot absorb/adjust without external help
6.	Describes short term fluctuations in weather variables	Describes long-term shifts in global or regional climate patterns

Table 1.1 *contd...*

S.No.	Climate variability	Climate change
7.	Temporary variations in weather/climatic conditions for a limited and specific area	Sustained changes in average climatic conditions over large areas
8.	The impacts are local, and regional within the earth's climate system Example: Drought, flood, etc.	Impacts are national and global Example: Raising global temperatures, Changes in global precipitation patterns
9.	Has a limited impact on production of farm and livestock	Has a major impact on production of farm and livestock
10.	Confined to short term affects which temporarily alter existing resource base	Long term effects which permanently alter the resource base
11.	Driven by fluctuations in temperature, precipitation, extreme weather events, etc.	Driven by natural, internal, external, and anthropogenic climate system

1.18 The Intergovernmental Panel on Climate Change (IPCC)

- The United Nations Environment Program (UNEP) and the World Meteorological Organization (WMO), established the Inter-Governmental Panel on Climate Change (IPCC) in 1988 to periodically assess the state of global environment and advise Members through UN agencies. Since then, the subjects of "Global warming" and "Climate Change" have drawn international attention
- The IPCC, brings together more than 2000 scientists and Government representatives to assess the risk posed by human induced changes in climate
- Its job is to assess the latest scientific, technical, and socio-economic literature on understanding the risk of climate change, its observed and projected impacts and options for adaptation and mitigation

- Since the creation of the IPCC, each Assessment Report has fed directly into international climate policymaking
- In 1990, the First IPCC Assessment Report (FAR) underlined the importance of climate change as a challenge with global consequences and requiring international cooperation. It played a decisive role in the creation of the UNFCCC, the key international treaty to reduce global warming and cope with the consequences of climate change
- The Second Assessment Report (SAR) (1995) provided important material for governments to the Kyoto Protocol in 1997
- The Third Assessment Report (TAR) (2001) focused attention on the impacts of climate change and the need for adaptation
- The Fourth Assessment Report (AR4) (2007) laid the groundwork for a post-Kyoto agreement, focusing on limiting warming to 2°C
- The Fifth Assessment Report (AR5) was finalized between 2013 and 2014. It provided scientific input into the Paris Agreement
- The IPCC has completed its sixth assessment cycle where it produces three Special Reports, a Methodology Report and the Sixth Assessment Report. The first of these Special Reports, Global Warming of 1.5°C (SR15), was requested by world governments under the Paris Agreement. In May 2019, the IPCC finalized the 2019 Refinement – an update to the 2006 IPCC Guidelines on National Greenhouse Gas Inventories. The Special Report on Climate Change and Land (SRCCL) was in August 2019 and the Special Report on the Ocean and Cryosphere in a Changing Climate (SROCC) in September 2019. The Sixth Assessment Report (AR6) was finalized in March 2023 in time for the first global stock take at the end of 2023
- In 2007, the IPCC and U.S. Vice-President Al Gore were jointly awarded the Nobel Peace Prize "For their efforts to build up and disseminate greater knowledge about man-made climate change, and to lay the foundations for the measures that are needed to counteract such change"

- In October 2022, the IPCC and the Intergovernmental Science-Policy Platform on Biodiversity and Ecosystem Services (IPBES) were declared co-laureates of the 2022 Gulbenkian Prize for Humanity
- The Jury of the Gulbenkian Prize for Humanity led by Dr. Angela Merkel, selected the IPCC and IPBES out of 116 nominations from 41 countries, in recognition of "…the role of science on the front line of tackling climate change and the loss of biodiversity"
- Inter Academy Council (IAC) conducts an independent review of IPCC procedures. The IAC issued a report and recommendations in August 2010. The IPCC considered these at its subsequent sessions, and made several revisions to its procedures as a result

1.19 Addressing Climate Change and Variability Challenges

- To address climate change and variability challenges individuals, families, communities, governments, etc., have the best sciences available to plan for and implement climate-smart and resilient practices. Two critically important approaches which are available to managing climate resilience are mitigation and adaptation. These strategies are interrelated and must be pursued simultaneously as both can potentially improve resilience to the changing climate and variability.

Chapter 2

Greenhouse Gases (GHGs) and Climate Change

"In Vedic thought the atmosphere is not merely empty space but a vibrant and essential element referred to as Vayu (air/wind). It is one of the five Mahabhutas (great elements) that constitute all existence both in the universe and within the human body. Vayu is the vital life force that sustains all beings from the simplest organism to the celestial bodies themselves. Its movement is responsible for essential bodily functions including respiration and circulation, and it is intrinsically linked to consciousness and spiritual energy"

2.1 Introduction to Greenhouse Gases (GHGs)

- Basics of Greenhouse Gases (GHGs)
  - The IPCC describes greenhouse gases (GHGs) as gaseous constituents of the atmosphere both natural and anthropogenic that absorb and emit radiation at specific wavelengths within the spectrum of radiation emitted by the Earth's surface, by the atmosphere itself, and by clouds
  - This property causes the greenhouse effect. Carbon dioxide, methane, nitrous oxide, water vapor (H_2O), and Ozone are the primary GHGs in the Earth's atmosphere
  - Human made GHGs include Sulphur hexafluoride (SF6), hydrofluorocarbons (HFCs), chlorofluorocarbons (CFCs)

and perfluorocarbons (PFCs). Several of these are also ozone depleting and are regulated under the Montreal Protocol

- The components of the climate system, viz., land, oceans, and atmosphere as well as living things affect atmospheric concentrations of green-house gases GHGs
- In the recent past, human influence has increased GHGs and has been the main cause of climate change
- In the industrial era, human activities added greenhouse gases to the atmosphere, and it is continuing
- Addition of more greenhouse gases such as carbon dioxide to the atmosphere intensifies the greenhouse effect thus warming the Earth's climate
- Continued emissions of other greenhouse gases viz., methane, nitrous oxide, hydrofluorocarbons, perfluorocarbons, etc., cause further warming and changes in all components of the climate system and threat agriculture, food production, and food security

- Carbon Cycle and Climate Change
  - The carbon cycle describes how carbon moves through Earth's atmosphere, land, and living organisms **(Fig. 2.1)**
  - This natural cycle helps regulate Earth's climate by controlling the amount of carbon dioxide (CO_2) in the atmosphere, a key greenhouse gas that traps heat from the sun
  - However, human activities, primarily the burning of fossil fuels (coal, oil, and gas), land use change, and deforestation are significantly disrupting this balance
  - By releasing large amounts of stored carbon into the atmosphere as CO_2, humans are amplifying the natural greenhouse effect, leading to an increase in global temperatures, a phenomenon known as global warming

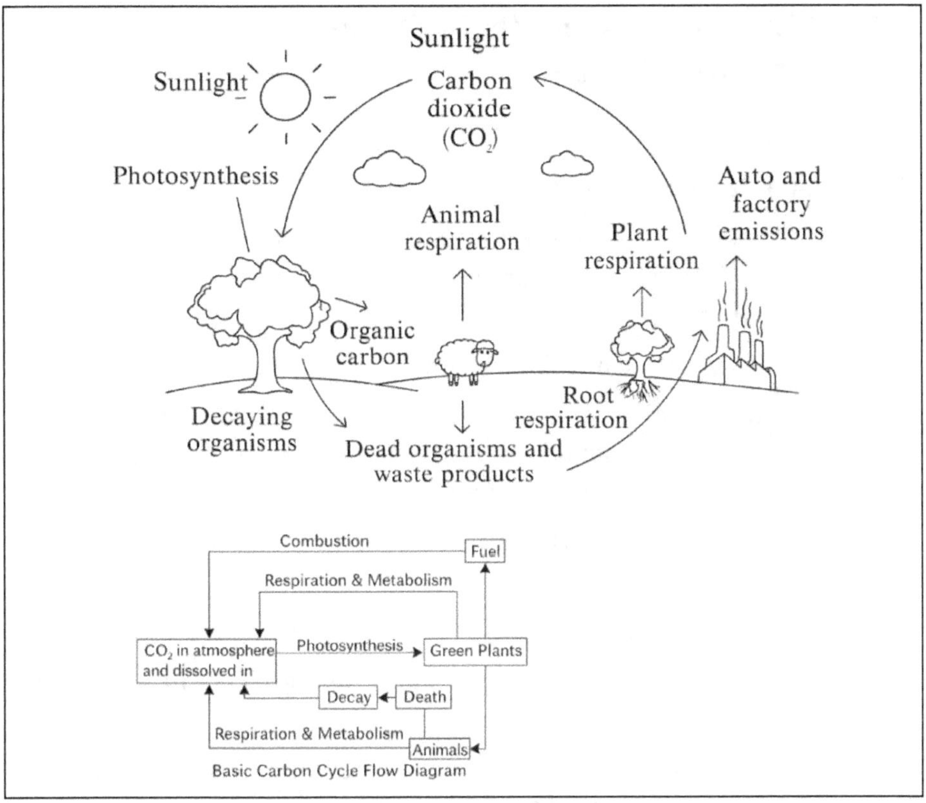

Fig. 2.1 The carbon cycle.
Source: Unknown. Class lecture notes collection.

- While natural sinks like oceans and forests absorb a portion of these human-caused emissions, their capacity to do so is finite and can be affected by climate change
- For instance, warmer oceans absorb less CO_2, and forests facing droughts or fires become less efficient carbon sinks
- Understanding and addressing the human impact on the carbon cycle is crucial for mitigating climate change and its associated consequences, such as more frequent and intense extreme weather events, rising sea levels, and ecosystem

2.2 Major GHGs and their Effects on Climate Change

2.2.1 Carbon Dioxide

2.2.1.1 Trends, Sources, and Projections

- Evidence from polar ice core samples show that the carbon dioxide level in the air has increased from 280 ppm in 1773 to 330 ppm in 1973
- However, the next increase of 50 ppm has taken only 33 years
- Some studies and estimates indicate that the CO_2 concentration increased at 1 ppm/year from 1960 to 1970 and this increase had been maintained at 1.5 ppm from 1970 to 2000
- In the decade 2001-2010 carbon dioxide emissions had been 2 ppm/year
- Currently CO_2 emissions are occurring at 3 ppm per year
- By the middle of this century plant usable CO_2 will increase by 50 per cent
- The normal value of CO_2 concentration assumed in agrometeorological studies is 330 ppm
- Decay of organic matter, industrial emissions, and burning of biomass release CO_2 into the air
- However, the major cause for sharp increase in CO_2 emissions is due to the burning of fossil fuels
- For CO_2 concentration to double in 2100 over the 2000 value, as envisaged by IPCC, the average annual emission level has to increase by 50 per cent compared to the current levels
- An increase in 25 per cent fossil fuel consumption or a decrease in the capacity of terrestrial and oceanic sinks for carbon dioxide can lead to this scenario
- Considering the history of CO_2 emissions and absorptions from the beginning of this century, the above possibility is very certain

2.2.1.2 Projections of IPCC on Carbon dioxide

- The IPCC (Intergovernmental Panel on Climate Change) has provided extensive assessments on Carbon Dioxide emissions, their current state, and future projections
- Atmospheric carbon dioxide levels indicate increasing trends **(Fig. 2.2)**

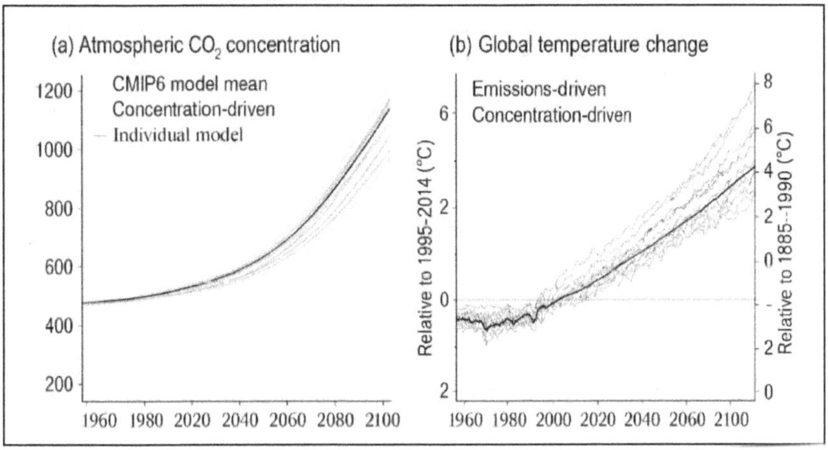

Fig. 2.2 Atmospheric carbon dioxide concentration (ppm) relative to 1995-2014.
Source: IPCC 2021, 2022, and 2023

- Human activities are the primary driver of increased atmospheric CO_2 concentrations, leading to global warming
- Atmospheric CO_2 levels are now 50% higher than pre-industrial levels and continue to rise rapidly
- The annual increase in CO_2 concentrations has accelerated, roughly doubling every 20-30 years over the last 60 years
- Diverse Scenarios: The IPCC analyzes various future scenarios, known as Shared Socioeconomic Pathways (SSPs), based on differing assumptions about population growth, economic activity, technology, and climate policies
- Paris Agreement Goal: Only the most optimistic scenario, involving a transition to sustainable practices and prioritizing well-being, aligns with the Paris Agreement's 1.5°C warming limit

2.2.1.3 Updated Atmospheric Carbon Dioxide levels at Mauna Loa Observatory

- According to updated Projections of NOAA, Scripps Institute of Oceanography, San Diego Atmospheric carbon dioxide levels at Mauna Loa Observatory also indicate increasing trends **(Fig.2.3)**

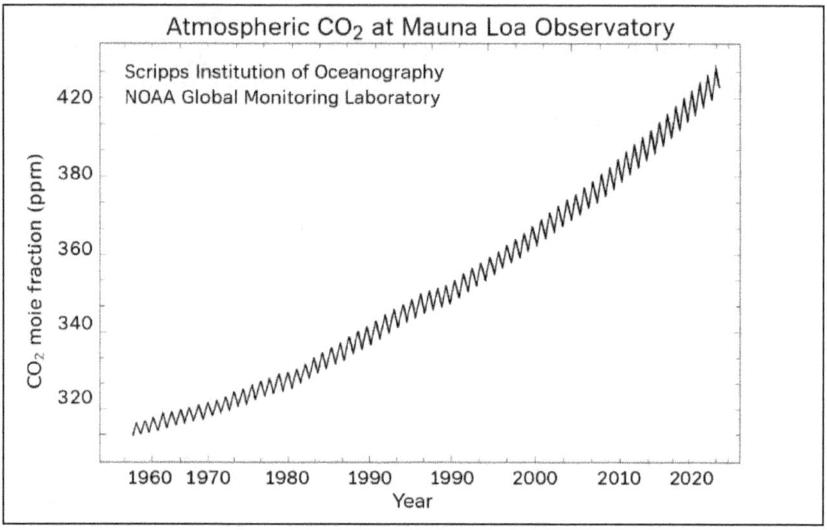

Fig. 2.3 Updated atmospheric carbon dioxide levels at Mauna Loa Observatory August, 2025.

2.2.1.4 General Effects of Carbon dioxide on Crops

- Elevated levels of CO_2 increase leaf area development
- Higher CO_2 levels also lead to a decrease in stomatal aperture which reduces transpiration
- The reduction in transpiration for a doubling of CO_2 ranges from 23 to 40 per cent
- The lowered stomatal conductance due to enhanced CO_2 does not interfere with gas exchanges between leaf and air and photosynthesis. Therefore, elevated CO_2 will increase photosynthesis

- Thus, elevated CO_2 helps crops to cope with drought conditions better
- The reduction in transpiration under elevated CO_2 increases leaf surface temperature there by crop canopy temperature
- It was found that the grain yields of many crops improve with an increase in canopy temperature above the ambient air temperatures. However, this increase is cardinal temperature dependent
- Research results indicate that on average a doubling of CO_2 will increase the yields of many crops by 30 per cent
- Elevated CO_2 increases dark respiration but the ratio of dark respiration to photosynthesis is constant under CO_2 enrichment
- Higher CO_2 benefits trees and shrubs more than grasses leading to woodlands expansion and encroachment of shrubs
- Elevated CO_2 increases photosynthetic rates and plant growth of natural ecosystems and of fruit trees like orange, citrus, apple etc.
- Elevated CO_2 is more beneficial for plants that fix atmospheric nitrogen
- Research in controlled environment conditions found that (Independent of temperature changes), for a doubling of CO_2 an increase in yield varies amongst crops being (a) 9 per cent for maize (b) 15 per cent and 28 per cent respectively for rice and wheat and 50 per cent for soybean
- It is reported that sour oranges exposed to CO_2 concentrations of 300 ppm over the normal produced 3 times more biomass than control (ambient carbon dioxide) in the first 2 years after which the rate of increase in biomass began to decline to a steady value of 1.7 times the control from the 9th year
- Since elevated CO_2 leads to an increase in biomass production with an accompanied decrease in water needs the water use efficiency for biomass production increases with an increase in CO_2 levels

- Elevated atmospheric CO_2 will have "Fertilization effect" increasing biomass and crop yield but necessarily not crop quality
- Wheat is a more efficient user of carbon dioxide than rice while maize is the least efficient user of carbon dioxide
- Under ideal conditions due to elevated carbon dioxide the photosynthesis increase by 30-50 per cent for C_3 crops (rice and wheat) and 10-25 per cent for C_4 crops (maize, sorghum, and pearl millet)
- An increase in carbon dioxide leads to a reduction in the quantum of outgoing longwave radiation and results in greater cloudiness which has effects on crops
- Doubling of Carbon Dioxide is expected to lead to an increase in mid and high-altitude cloudiness
- Carbon dioxide is the source of carbon for photo synthetically derived carbohydrates and is therefore of crucial importance for food security
- CO_2 is also important in the longwave radiation transfer characteristics of the atmosphere
- It absorbs and therefore emits radiation at several wave lengths viz., 1.6, 2.0, 2.7, 4.3 and 15 microns
- The major sinks for CO_2 are terrestrial and aquatic vegetation during periods of active photosynthesis
- CO_2 production from the soil leads to high CO_2 concentration at the base of the canopy
- At night the canopy releases respiratory CO_2 and the concentration increases rapidly downward through the canopy, reaching a maximum at ground level. This helps with more photosynthesis in the early hours of the day
- C_4 crops (maize, sorghum, pearl millet) require less water per gram of carbon assimilated than C_3 crops (rice, wheat)
- Therefore, sorghum and maize are preferred in dry land areas

2.2.1.5 General Effects of Carbon Dioxide on Pests and Diseases

- Elevated CO_2 effects occurrence of pests and diseases of crops
- Higher CO_2 favor pathogens because of heavier foliage development
- Longer and increased availability of crop residues on account of their lower rate of decomposition under elevated CO_2 help in easier over-wintering of pathogens favoring earlier and rapid development of diseases
- Increase in levels of simple sugars in the leaves of soybean under elevated CO_2 is a contributory factor for increased damage by many insects
- The lowered nitrogen content of leaves under elevated CO_2 leads to voracious consumption of leaves by insects to meet their metabolic requirements

2.2.1.6 Role of Agriculture and Forestry in Mitigation of Elevated Carbon Dioxide

- Vegetation is both sink and source for CO_2
- Atmospheric CO_2 can be returned to the land by afforestation, conservation tillage by a cover crop, and no-till conservation practices
- Forests cover 1/3 of the land area and contribute to 2/3 of global photosynthesis
- Thus, forests constitute a major terrestrial sink for CO_2 as important as that of the oceans
- Deforestation contributes in a negative way to build up of CO_2 in air
- The scope for reducing CO_2 emissions from crops and agricultural fields is very limited
- This makes an immediate need for increasing the sink capacity for absorption of CO_2
- Increasing (a) the field-occupancy of crops in time and/or space, (b) area under perennial vegetation, and (c) per day unit area yields of crops would constitute additions to sink capacity for CO_2

- These can be achieved through technological interventions like crop breeding, optimal use of surface irrigation, social forestry, agroforestry etc.

2.2.1.7 Carbon Dioxide and Carbon Sequestration

- Carbon sequestration is both a natural and anthropogenic process by which CO_2 is diverted from sources to sinks
- Soil is the bio-geo-chemical interface among the atmosphere, vegetation, and hydrosphere which is actively involved in the cycling of atmospheric and terrestrial carbon
- Thus, besides vegetation, soil is a significant carbon storage sink
- The conversion of carbon dioxide by vegetation into biomass constitutes direct carbon sequestration
- In soils carbon sequestration occurs due to chemical reactions that convert CO_2 in organic carbon to carbonates
- The carbon in the biomass stored as wood in trees can remain sequestered for (i) centuries if used in a non-destructive way such as making furniture, doors, windows, etc., and (ii) for decades if used as fuel
- Burning of biomass from vegetation is a significant source of emission of CO_2
- However, if the biomass is converted into a value-added product like compost and returned to the soil, it is stored as Soil Organic Carbon (SOC), and the carbon remains without immediate remission
- Soil is also a source of CO_2 and over the years much SOC has been lost due to a variety of reasons
- Thus, SOC shows cycles of accumulation and loss and attains a quasi-equilibrium state over a period which depends on management practices, soil, and climate
- Loss of SOC is reversible. However, pushing up the SOCs successively to higher quasi equilibrium levels calls for changes in land-use-pattern, vegetation cover, and management practices

- The capacity of soil for sequestering carbon is only about one-tenth of that of vegetation
- Current potential for sequestration of carbon in form of SOC by following recommended management practices for agricultural soils is held to be one-fourth to one-third of annual increase in atmospheric CO_2
- Smectites (swelling clays) improve soil carbon sequestration and in semi-arid tropics with Smectite rich soils a nearly fourfold increase in SOC in the last 25 years has been reported
- In implementing SOC enhancement projects the trade-off between various competing interests has also to be kept in view

2.2.2 Methane

2.2.2.1 Trends, Sources, and Projections

- The concentration of methane, which was 700 parts per billion (ppb) in 1750 had increased to 1.75 parts per million (ppm) in 1990
- This is a 250 per cent rise compared to a 25 per cent rise in carbon dioxide
- Methane constitutes only 2 ppm of air in 2000
- A methane molecule is 30 times more efficient in trapping of heat compared to a molecule of CO_2
- Methane emission slowed down in the 1990s to remain at a steady value of 1.75 ppm from 1999 to 2002 and had increased by only 4 per cent from 1990 to 2000
- Unlike carbon dioxide which remains in the atmosphere for over 100 years methane has a half-life of only 8 to 12 years in the atmosphere
- Animals, especially livestock production significantly contributes to methane emissions

- Ruminant animals like cattle, sheep, and goats possess a unique digestive system that involves fermentation of food in their rumen
- This process releases methane as a byproduct primarily through burping
- Enteric fermentation accounts for the largest portion of methane emissions from animals estimated at 73% of agricultural methane emissions or 88% of livestock-related greenhouse gas emissions worldwide
- Rice paddies, production and use of coal, natural gas, oil, and decay of organic matter in municipal solid waste landfills, are the other major sources for emission of Methane
- Rice paddies contribute 25 per cent of the total global production of methane while the contribution from livestock is 16 per cent
- Wetland rice as a major source of atmospheric methane has come into light in the early 1980s
- Puddling of rice fields leads to cutoff of oxygen supply from the atmosphere to the soil resulting in anaerobic fermentation of organic matter by methanogenic bacteria and in production of methane, by trans-methylation of acetic acid and reduction of CO_2
- Methane remains trapped in an undisturbed flooded paddy field with only a small amount escaping into the air as bubbles
- Cultural practices associated with irrigated lowland rice account for 30 per cent of the soil methane released into the air
- Aerenchyma cells of the rice plant provide the conduit for 70 per cent of the Methane released into the air
- Higher biomass production of rice and higher incorporation of organic matter to puddle fields increase methane emissions
- When manure is stored particularly in liquid form in lagoons or holding tanks it decomposes in the absence of oxygen (anaerobically) leading to the release of significant amounts of methane
- This accounts for about 7% of agricultural methane emissions

2.2.2.2 Reducing the Methane Emissions

- Breeding higher quality cattle and improved feeds reduce CH_4 emissions from domestic livestock
- Even medium-strong measures reduce methane emission by $1/8^{th}$
- Therefore, the concentration of methane in air in 2050 will be 3.6 ppm
- For example, rice cultivation under flooded conditions is a major source of atmospheric methane
- *In situ* reduction of methane from flooded rice fields is a must
- Switch over from flooded paddies to a new and recent system of rice cultivation i. e., the SRI (System of Rice Intensification) lowers about 20 per cent of global warming potential (GWP) compared to flooded rice
- Draining of puddled rice once or several times during the growing season has been reported to reduce methane emissions
- Of all forms and permutations of draining of water aeration provided by multiple drainage emerges as the most effective in reducing methane emissions
- Pure mineral fertilizers and pure non fermented organic fertilizers produce the lowest and highest emissions of methane respectively
- Amendments that reduce methane emissions become important
- Use of Calcium Carbide and Dicyanamide to reduce emission of CH_4 in rice is noteworthy
- Cultivars with short root systems, high root oxidative activity, high harvest indices, and more tillers are likely to produce less methane than other cultivars
- Compared to transplanted rice, direct seeding of pre-germinated rice has been reported to reduce methane emissions on account of reduced flooding periods and little soil disturbance

- As SRI results in considerable saving of irrigation water the saved water may be used to increase the area of other crops
- Soil microorganisms remove methane with improved pasture conditions and cover crops on cultivated land
- As disturbance of soil leads to emission of GHGs from soil and in view of the advances made in weed control and farm machinery, the concept of minimal or no tillage is being advocated as a mitigation measure against emission of GHGs from soil in general and methane in particular

2.2.3 Nitrous Oxide

2.2.3.1 Trends, Sources, and Projections

- The concentration of nitrous oxide has steadily increased from a value of 270 parts per billion (ppb) in 1750 to 314 ppb by 1998
- 90 per cent of nitrous oxide in the atmosphere is of agricultural origin
- Therefore, the increase in nitrous oxide up to 2020 was 15 per cent and increase from 2020 to 2050 will be 22.5 per cent
- Thus, the concentration of nitrous oxide in air in 2050 will be 330 ppb
- A nitrous oxide molecule is 270 times more effective in trapping heat than a carbon dioxide molecule
- Like carbon dioxide, nitrous oxide remains in the air for about 100 years
- A highly significant source of emission of nitrous oxide into the atmosphere from cropped fields is the application of inorganic nitrogenous fertilizers and other nitrogenous materials more than the nitrogen requirements of crops
- The nitrous oxide emissions arise from nitrification of mineralized organic nitrogen in the topsoil and denitrification of inorganic nitrogen in the wet subsoil

2.2.3.2 Reducing the Nitrous Oxide Emissions

- Even medium-strong measures reduce nitrous oxide emissions by 15 per cent
- The mitigation options include practicing precision farming in which the rates and timing of nitrogenous fertilizers are so adjusted as to meet the exact nitrogen needs of crops
- Adopting practices that deliver nitrogen more efficiently to crops
- Using fertilizers that release nitrogen in a slow and controlled manner
- Proper methods of application of fertilizers
- Use of nitrification inhibitors
- Use of amendments that reduce nitrous oxide emissions like Calcium Carbide and Dicyanamide in rice crop
- Soil microorganisms remove nitrous oxide with improved pasture conditions and cover crops to lower the amount of inorganic nitrogen in the soil

2.2.4 Water Vapor

2.2.4.1 Role of Water vapor in regulating Earth's Temperature

- Water vapor is a greenhouse gas and plays a significant role in regulating the Earth's temperature
- Water vapor, like other greenhouse gases, absorbs infrared radiation (Heat) emitted from Earth's surface and reradiates it back down, warming the planet
- As a feedback mechanism, when other greenhouse gases (CO_2) cause an initial warming, more water evaporates and enters the atmosphere
- This increased water vapor then absorbs more heat, causing further warming, creating a positive feedback loop

- Essentially, warmer temperatures lead to increased evaporation, resulting in more water vapor in the atmosphere, which further traps heat and amplifies the warming effect
- Increased water vapor feedback amplifies the direct radiative effect of carbon dioxide
- Carbon dioxide absorbs energy in some small segments of the thermal infrared segment that water vapor misses and warms air a bit more
- Doubling of carbon dioxide is held to be equivalent to a 13 per cent increase in water vapor
- Increase in temperature increases the capacity of air to hold water in vapor form
- However, continued increase in water vapor leads to formation of clouds which will reduce radiation receipt and exert a cooling effect
- Which of the two processes will eventually dominate and when and where are difficult to predict as of now with the available knowledge
- Thinning of low clouds in subtropics and mid-latitudes with lower albedos due to warming has been indicated
- Water vapor concentrations fluctuate regionally due to weather patterns, but human activities result in localized effects, such as increased evaporation from irrigation

2.2.4.2 Trends, Sources, and Projections

- Atmospheric water vapor, a key greenhouse gas, exhibits increasing trends and is projected to further rise in the future because of human-induced climate change
- Since the 1980s, studies consistently show a rise in global atmospheric water vapor with a global average increase of 1-2% per decade
- Regional variations are also observed, particularly in areas like South Asia where positive trends in water vapor have been identified

- The Arctic region shows significant positive trends in water vapor, likely due to rapid warming leading to increased evaporation
- Despite the overall global increase, the rise in near-surface water vapor pressure over land and ocean has weakened since the late 1990s in certain regions
- Climate model projections for future high-emission scenarios predict a substantial increase in atmospheric water vapor, particularly in polar latitudes, potentially doubling current levels by the end of the 21st century
- This rise in water vapor could further exacerbate global warming and phenomena like Arctic amplification impacting global and regional climates
- Under high-emission scenarios, the lifetime of water vapor could increase by 25% by the end of the 21st century
- This indicates important changes in the hydrological cycle, with potential implications for precipitation patterns and extremes

2.2.5 Ozone

2.2.5.1 Ozone as Greenhouse Gas

- Ozone is a greenhouse gas, but it differs from other greenhouse gases in several ways
- The effects of ozone depend on its altitude, or where the gas is located vertically in the atmosphere
- Most ozone naturally exists in the layer of the atmosphere called the ozonosphere
- Ozone in this part of the atmosphere has a slight net warming effect on the planet, but it is good for life on Earth, because it absorbs harmful ultraviolet radiation from the sun preventing it from reaching the Earth's surface
- In the troposphere (layer of the atmosphere near ground level) ozone is an air pollutant that is harmful to breathing, a main ingredient of urban smog and an important greenhouse gas that contributes to local climate change

- Unlike the other major greenhouse gases tropospheric ozone only lasts for a few days to weeks and the levels often vary by location

2.2.5.2 Trends, Sources, and Projections

- Globally, the ground-level ozone, i.e., tropospheric ozone increased from approximately 47 ppb in 2010 to almost 50 ppb in 2019
- Significant increases have been observed in regions like East Asia, the Persian Gulf, India, northern South America, the Gulf of Guinea, and Malaysia/Indonesia
- In megacities, ozone levels between 2005 and 2019 showed a significant global increase, with anomalies reaching up to 5 ppb in certain areas
- The ozone layer is recovering in ozonosphere/stratosphere due to the phasing out of ozone-depleting substances (ODSs) under the Montreal Protocol
- The Antarctic ozone hole is shrinking, and stratospheric ozone is expected to return to 1980 levels by around 2066 in the Antarctic, 2045 in the Arctic, and 2040 for the rest of the world
- Climate change is projected to increase tropospheric ozone concentrations, particularly in high-income countries, if precursor emissions remain constant
- Studies predict a potential increase in daily maximum ozone concentrations due to the higher probability of heatwaves, increased solar radiation, and faster chemical processes of ozone formation
- The future trajectory of ozone concentrations is closely tied to emissions of precursors like nitrogen oxides

2.2.5.3 Ozone and Climate Change

- Increased concentrations of tropospheric ozone contribute to global warming and its associated impacts (rising temperatures, extreme weather, etc.)

- Ozone interacts with other greenhouse gases like carbon dioxide and methane influencing the overall warming effect
- While ozone depletion in the stratosphere has a cooling effect, the increase in tropospheric ozone due to pollution has a warming effect

2.2.5.4 Importance of Reducing Tropospheric Ozone

- Tropospheric ozone is also a respiratory irritant and can cause significant health problems
- It can also damage plants impacting agricultural yields and ecosystems
- Reducing emissions of ozone precursors is crucial for agriculture, human health, and climate change mitigation

2.2.6 Fluorocarbons

2.2.6.1 Fluorinated Carbons as Greenhouse Gases

- Fluorinated gases, including Hydrofluorocarbons and Perfluorocarbons act similar to other greenhouse gases by absorbing and trapping infrared radiation (heat) that radiates from the Earth's surface
- Fluorinated gases have a disproportionately large effect on the greenhouse effect due to their high Global Warming Potential (GWP) compared to other gases like carbon dioxide
    - GWP is a measure of a greenhouse gas's heat-trapping ability compared to a ton of CO_2 over a given time frame (typically 100 years)
    - HFCs and PFCs have GWPs thousands of times greater than CO_2. For example, HFC-23 has a GWP that is 14,800 times higher than CO_2 over 100 years
- These gases persist in the atmosphere for a significant period from several years to centuries in case of some HFCs, and even thousands to tens of thousands of years for PFCs
- The prolonged presence in the atmosphere results in continuous warming impact for an extended time
- For instance, tetrafluoromethane, a type of fluorocarbon, has an atmospheric lifetime of 50,000 years

- By trapping heat and lingering in the atmosphere for extended periods, fluorinated gases contribute to a warming of the Earth's surface, amplifying the greenhouse effect and resulting in climate change

2.2.6.2 Trends, Sources, and Projections

- The HFCs and PFCs are entirely human-made compounds, not naturally present in the atmosphere
- The HFCs were initially introduced as replacements for ozone-depleting substances (ODS) like chlorofluorocarbons (CFCs) and hydrochlorofluorocarbons (HCFCs)
- They are widely used as refrigerants in air conditioning, refrigeration, foams, aerosols, fire protection systems, and solvents
- The PFCs are produced as a byproduct of aluminum production and are used in semiconductor manufacturing
- The emissions of these fluorinated gases are regulated and controlled by Montreal Protocol
- The Montreal Protocol successfully led to a significant reduction in ozone-depleting substances, showcasing the power of collective commitments to environmental health
- Although emitted in smaller quantities compared to CO_2, the potent heat-trapping ability and long atmospheric lifetimes of fluorinated gases make them significant contributors to global warming and climate change
- However, regulated and controlled by Montreal Protocol these are under reducing trends

2.3 Addressing GHG Emission Challenges

- The individuals, families, communities, governments, organizations etc., have the best sciences available to plan for and implement climate-smart and resilient practices under enhanced GHG emissions. Two critically important strategies available are mitigation and adaptation. These strategies are interrelated and must be pursued simultaneously as both can potentially improve resilience to the changing climate due to GHG emissions.

Chapter 3

Climate Change and Solar Radiation

"In the Vedas Solar Radiation is profoundly depicted as not merely physical energy, but the very essence of life and consciousness. The Divine Sun (Surya) illuminates the external world and ignites the inner spirit (Atman), guiding towards enlightenment and self-realization. Thus, the radiant energy nourishes creation, inspires wisdom, and purifies the soul connecting all existence to the boundless source of light"

3.1 Introduction to Solar Radiation

- The energy received from the sun is known as "Solar radiation"
- The sun emits radiation over the whole of the electromagnetic spectrum from gamma rays to radio waves
- Solar radiation controls life by heating the earth and atmosphere and by providing the energy required in photosynthesis for conversion of carbon dioxide and water into the primary sources of food "Carbohydrates"
- The greater part (98 per cent) of this radiation is emitted in the waveband 0.25 – 3.0 microns
- The human eye can detect solar radiation in the waveband 0.4-0.7 microns
- This process is called vision and solar radiation in this waveband when concerned with vision is called "Light"

- Discrete particles of radiation are termed "Quanta" and quantum of light is called a "Photon"
- The general term applied to the radiant flux density incident on a unit surface is "Irradiance"
- All materials on earth absorb and emit radiation at longer wavelengths than solar radiation approximately between 3 and 100 microns
- The flux of the emitted longwave (infrared) radiation is proportional to the temperature of the body and its emissivity
- Most natural objects absorb nearly all the incident radiation in the waveband from 3 to 100 microns and are called "Black bodies" in this waveband
- All good absorbers are good emitters of radiation and efficiently emit radiation in this waveband
- A terrestrial body absorbs and emits longwave radiation even when solar radiation is absent
- Longwave radiation transfer is important in determining the thermal or temperature relationships of organisms and their environment. Example: The walls of the room exchange longwave radiation with people in the room. In the absence of a large influx of solar radiation, longwave radiation exchange between bodies of people in the room and the walls determines the level of comfort
- The atmosphere is differentially transparent to longwave (terrestrial/infrared) and shortwave radiation (solar radiation), being highly transparent to shortwave radiation and partly opaque to longwave radiation
- The opacity of the atmosphere to longwave radiation is due to the absorptivity of water vapor, liquid water, carbon dioxide, etc., for longwave radiation
- However, at wavelengths between 10 and 12 microns the absorptivity of water and carbon dioxide is very low

- Therefore, the atmosphere is effectively a "Window" for longwave radiation in this wave band
- In the clear atmosphere the temperatures are as low as -80 degrees centigrade in the troposphere and the atmosphere represents a sink for longwave radiation emitted by warmer terrestrial organisms
- The waveband of the window also coincides with the peak emission of many terrestrial organisms and so there will be a longwave flux from the organisms through this window
- The influence of the window can be felt by humans on clear nights which feel cold, in contrast to cloudy nights which even at the same air temperature feel warmer
- On cloudy nights the clouds are at higher temperatures than the cloudless atmosphere and emit a greater downward flux

3.2 Trends, Sources, and Projections

- Theoretically, enhanced carbon dioxide should lead to a decrease in mean daily global solar radiation
- Reduction in solar radiation at 2 per cent per decade at many stations in India during the period of May to December is reported
- It is also reported that in many industrial regions of the globe solar radiation has decreased by 2.7 per cent per decade during the past 50 years
- This is due to an increase in cloudiness because of anthropogenic aerosols and pollutants changing the optical properties of the atmosphere
- A reduction in solar radiation normally equals 2/3rds of the increase in cloudiness
- Thus, cloudiness is expected to increase by 20 per cent by 2050

3.3 General Effects of Solar Radiation on Crops

- Crop yield is a result of photosynthetic production and respiratory consumption of dry matter and apportioning of the net accumulated dry matter to the economic yield components of the crop

- Plants with higher leaf area intercept more solar radiation ultimately contributing to improved crop yields
- Wide differences exist among crops in their responses when exposed to the same weather and the same holds good for photosynthesis also
- The response of crops to weather elements is affected by the growth stage and physiology of crops
- Photosynthesis is better when the Sun's rays are slanted
- More the "Photo period" more is the "Photosynthesis"
- Therefore, there are differences in the yields of crops across latitudes with better yields at higher latitudes than lower latitudes
    - Maize, tobacco and soybean are the most efficient, least efficient, and average users of PAR respectively
    - Reduced solar radiation in the vegetative phase of rice has little influence on yield except under excessively cloudy conditions
    - However, solar radiation from panicle initiation to maturity is linearly related to the yield of rice
- Reduction of bright sunshine hours during specified crop stages of growth of several crops reduces crop yields. Example: Reduction in sunlight in the six (6) week period following panicle initiation in rice reduces grain number, increase spikelet sterility, and ill filled spikelets resulting in lower rice yields
- The net primary production is proportional to the Intercepted Photosynthetically Active Radiation (IPAR)
- Therefore, maximization of interception of solar radiation by the crop canopy in its growing period and efficient use of available solar radiation are required to bring about a quick increase in productivity of crops
- Agronomic strategies must ensure that the crops develop to reach ground shading stage in the quickest possible time with minimum possible number of leaves per plant to avoid mutual shading which lowers photosynthetic rates of shaded leaves

- To achieve this, adoption of proper sowing window, crop row orientation, crop geometry, and choice of proper crop variety are necessary
- Crop duration of many crops, particularly groundnut is mainly determined by the duration of vegetative and reproductive stages
- In addition, for intercropping and mixed cropping plant competition for PAR may change as it influences the availability of carbon dioxide, nutrient, water, etc.
- Therefore, in mixed cropping system, the taller crop should be of the C_4 type and the shorter one of the C_3 type to achieve optimum crop yields

3.4 Absorptivity of Individual Leaves

- In the Photosynthetically Active Radiation (PAR) region (0.4 - 0.7 microns) the leaf absorbs 90 per cent of the radiation incident upon it
- In the shortwave infrared (0.7 to 3.0 microns) it avoids a large heat burden by reflecting and transmitting most of the radiation, which the leaf does not use
- In the longwave radiation waveband (3-100 microns) the leaf strongly absorbs 90 per cent of radiation incident upon because the leaf needs radiant energy to drive its photochemistry
- The leaf which absorbs strongly in the longwave IR (3-100 microns) there is no major input from the Sun
- As per Kirchoff's Law leaf is a good radiator in this waveband and dissipates upward energy as heat very efficiently over the longwave radiation waveband
- Approximately half of the shortwave radiation falls into the visible or Photosynthetically Active Radiation (PAR) region of the spectrum (0.4-0.7 microns) and the other half in the shortwave infrared (0.7-3.0 microns)

- The importance of the high absorptivity of the leaf in the longwave region is due to the correlation between emissivity and absorptivity as described by Kirchoff's law
- Thus, bodies with high absorptivity have high emissivity and vice versa
- At a temperature of 293 °K (20 °C) the leaf has a peak emissivity at 9.89 microns which is in the longwave window of atmosphere
- Also, the emitted longwave flux from the leaf is large, with a typical value of 390 wm^{-2} at 293 °K (20 °C) indicating the importance of the longwave flux in reducing the energy load and therefore the heating effect on the leaf
- Because of this reason the optimum temperature for maximum performance of crops is fixed at 20 degrees centigrade
- On a clear sunny day the leaf absorbs solar radiation in the 0.4-0.7 microns region
- The longwave radiant flux increases with temperature approximately 30 per cent from 283 °K (10 °C) to 303 °K (30 °C) and emitted from the leaf to the atmosphere, thereby dissipating a large portion of heating effect of the absorbed solar radiation

3.5 PAR (Light) and Crop Growth

- The influence of light on plants is complex
- It drives the photochemistry of the plant and affects the shape and size of the plant
- Plants change their strategy of assimilate utilization in response to changes in the energy received from light as well as changes in the rates of photosynthesis
- Light quality particularly the ratio of red (660 microns) to far red (730 microns) affects the growth and development of the plant
- The ratio of red to far red light in the natural environment is significant for crop growth and development
- Above a solar elevation of approximately 10 degrees the ratio remains constant at about 1:1
- However, at sunrise and sunset rapid changes occur in the ratio

- This is the reason for differential absorptions of solar radiation during different periods of the day by plants
- Usually more both during morning and evening because the sun's rays are slanty during these hours
- Because of the spectral absorption characteristics of leaves there are large changes in the ratio within crops and particularly between shade and sun flek areas
- The periodic nature of irradiance influences the vegetative and reproductive stages of a plant

3.6 Intercepted Solar Radiation and Time of Growing Season

- Percentage of solar energy intercepted by crop canopy is crucial for biomass/dry weight production and yield
- Plants utilize solar energy through photosynthesis to convert light energy, carbon dioxide, and water into organic matter, forming the basis for growth and development
- The intercepted and converted energy make up for dry weight
- As the time in growing season increases both the solar energy intercepted and converted energy increases steadily **(Fig: 3.1)**

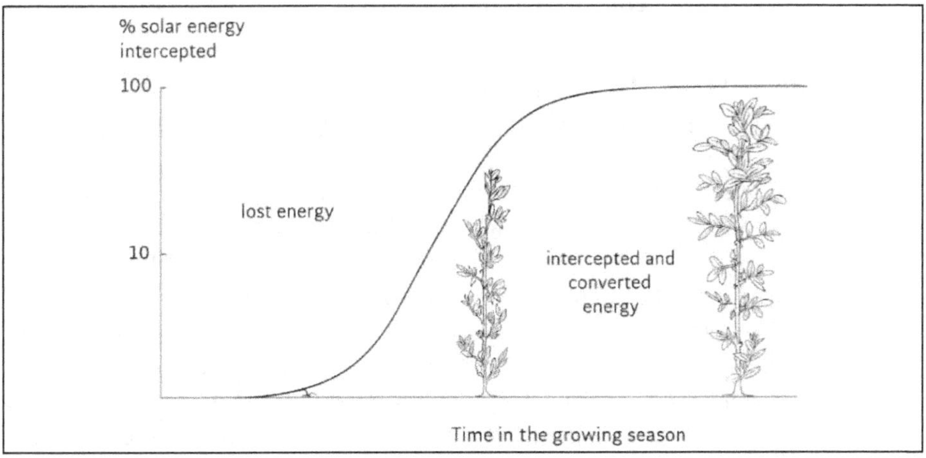

Fig. 3.1 Relationship between intercepted solar radiation and time of growing season.

Source: www.livingfield.co.uk

3.6.1 Factors influencing Intercepted PAR and Crop Growth

- Planting density and canopy structure: Optimized planting density and plant architecture, including leaf angle and distribution, significantly affect light interception and penetration into the canopy. A well-designed canopy ensures efficient light capture, particularly in the lower canopy layers, which can significantly enhance crop yield
- Leaf Area Index (LAI): LAI, representing the total leaf area per unit ground area, directly impacts light interception. Optimizing LAI is crucial for maximizing light capture while minimizing self-shading, particularly in dense canopies
- Light Use Efficiency (LUE) or Radiation Use Efficiency (RUE): LUE/RUE is a measure of the efficiency with which plants convert intercepted PAR into biomass. It is a critical factor influencing crop yield
- Environmental and other factors: Factors like temperature, water availability, nutrient availability, carbon dioxide levels, and even atmospheric conditions like wildfire smoke and cloud cover can impact light interception and, consequently, crop growth and yield
- Effectively intercepting and efficiently utilizing solar radiation is fundamental to maximizing crop growth and yield potential

3.7 Basic Micrometeorological Terminology and Laws of Solar Radiation for Climate Change Studies

- Fewer than 5 per cent of photons absorbed by leaves are used for photosynthesis. That is why the Radiation Use Efficiency (RUE) of most of the crops is around 5. In addition, most of the photons that are not used in photosynthesis tend to heat the leaf. However, the temperature of the leaf does not increase continuously throughout the day. The energy of the excess photons is dissipated in transpiration, convection, and radiation. To partition this energy between the various processes one must understand basic laws

- First law of thermodynamics: Energy can neither be created nor destroyed i.e., the energy that enters a crop canopy must be equal to the energy stored by the crop plus energy leaving it
- Kirchoff's law: A good absorber of radiation is a good emitter, in similar circumstances. The absorptivity "a" of an object for radiation of a specific wavelength is equal to its emissivity "e" for the same wavelength
- Stefan Boltzmann law: The radiant energy emitted by a body is a function of the 4th power of its temperature in Kelvins
- Wein's Displacement law: The wavelength at which the maximum amount of radiation is emitted decreases as the temperature of the body increases
- Thermal radiation: All the bodies continuously emit and absorb longwave radiation and that which is not absorbed is reflected. The emission of radiation is a consequence of the excitation of the atomic and molecular structure and is a function of temperature
- Short and long wavelengths: The wavelengths at which radiation is emitted by a body at terrestrial temperatures are very much greater than those emitted by the sun. Thus, solar radiation is often called "Shortwave" and "Terrestrial/ Thermal radiation/Infrared" radiation is often called "Longwave"
- Radiant flux: The amount of radiant energy received, emitted, and transmitted per unit time
- Radiant flux density: Radiant flux per unit area of a surface oriented right angles to the source
- Solar radiation flux is strongly influenced by the atmosphere around the earth

3.8 Solar Radiation Techniques for Climate Change Mitigation and Adaptation

3.8.1 Effects of Surface Geometry on Solar Radiation

- The effects of small terrace irregularities such as mounts, ridges, furrows, trenches, etc., alter the amount of solar radiation (in particular the direct solar radiation) striking the surface thereby soil temperature and soil moisture are altered

- The germination and emergence of a considerable range of crops thereby changes
- Day time geometrical relationships between solar radiation and slope and the aspect of small ridges and furrows depend on their orientation with respect to slopes and position of the sun
- In the early morning the tops of ridges are colder than at ground level, both because of excessive sky view and because of the relatively large distance to deep soil heat
- Thereafter, sunny slopes heat up appreciably, depending on orientation, also because reflected radiation may still be absorbed at the other side (radiation trapping)
- An interesting situation occurs on very dry days as follows
    - North-South oriented ridges where plants placed on west-facing slopes survive because in the early shade they make use of occasional morning dew
    - Nocturnal temperature relationships in small hollows and trenches (frost hollows) help to reduce the soil temperature at micro level in summer season. Therefore, vegetables such as tomatoes, chilies, etc., are grown successfully in summers

3.8.2 Relation of Sun Angle to Effectiveness of Insolation

- The effectiveness of insolation (incoming solar radiation) is directly related to the angle at which the sun's rays strike a surface **(Fig. 3.2)**.
- Higher angle (more direct sunlight): When sunlight hits a surface at a higher, more perpendicular angle (closer to 90 degrees), the energy is concentrated over a smaller area, leading to greater intensity and more effective heating. This happens, for example, near the equator at noon locally
- Lower angle (more slanted sunlight): When sunlight hits at a lower, more oblique angle, the same amount of energy is spread out over a larger area, resulting in lower intensity and less effective heating. This occurs at higher latitudes
- This principle is crucial in understanding Earth's climate zones. Equatorial regions are warmer due to more direct sunlight

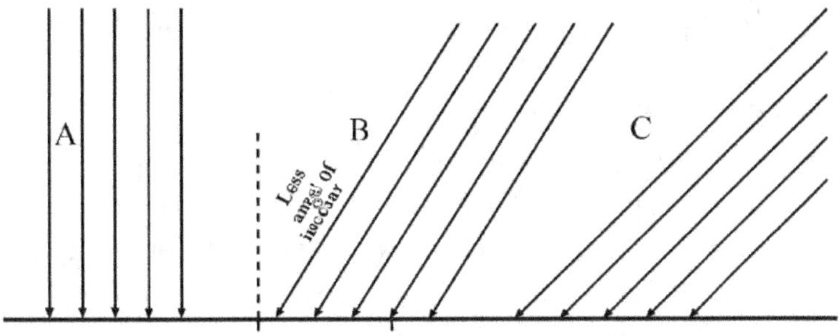

A. Sun Directly Overhead (Less spread of energy)
B. and C. Greater Angles of Incidence (More spread of energy)

Fig. 3.2 Relation of Sun angle to effectiveness of insolation.
Source: Unknown. Class lecture notes collection.

- On the same analogy, in crop canopies when sunlight hits at a lower, more oblique angle, the amount of energy is spread out over a larger leaf area of crop both in morning and evening resulting in more biomass production. This is more so in tropical areas where the intensity of solar radiation is more in all seasons

3.8.3 Crop Management and Layout

- Crops can be raised more successfully, and profits could be increased by covering the soil by changing its surface shape as compared to crop raising under natural conditions
- When the crop covers the ground completely, the absolute value of temperature, humidity, and winds at the surface and the vertical gradients of these elements are decisively altered by the crop itself
- On horizontal ground the spacing of row crops and the seasonal changes in the ratio of crop height to distance between the rows drastically affect the amount of sunshine reaching the actual soil surface (radiation phenomena depend upon the ratio of direct and diffuse sunshine)
- Crop yield per unit area is affected by both the density of planting and the production of individual plants

- If densities are higher than optimal then the light which can still penetrate to lower parts of the crop suffers too much intensity reduction in PAR part of the spectrum
- In addition, movement of air within the crop decreases with increasing density which may reduce the CO_2 concentration near lower leaves and limit photosynthesis on sunny days with weak winds. Density of rows (spacing) may have little effect if sufficient water is available
- However, in arid conditions wide spacing is recommended

3.9 Addressing Climate Change Challenges of Solar Radiation

- Individuals, families, communities, governments, organizations, etc., have the best sciences available to plan for and implement climate-smart and resilient practices. Two critically important approaches which are available to managing climate resilience are mitigation and adaptation. These strategies are inter-related and must be pursued simultaneously as both can potentially improve resilience to the changing climate.

Chapter 4

Climate Change and Wind

"Vayu, the Breath of Brahman pervades all existence sustaining life and symbolizing the omnipresent and unseen force that binds the universe. As Prana (life-breath) it embodies vital energy within each being and controlling it through practices like Yoga and Pranayama leads to supreme wisdom and inner harmony. Vayu represents both the dynamic forces of creation and destruction highlighting the inter-connectedness of all things and the constant flow of transformation in the cosmos"

4.1 Introduction to Wind

- Air in its horizontal motion is known as "Wind"
- Air in its vertical motion is known as "Air current"
- Wind near the ground is very unsteady
- Gusty conditions are frequent near ground representing large scale turbulence
- The more unstable the atmosphere the more the turbulence
- It is because of enhanced mixing
- It is the turbulence which is the primary factor determining aerodynamic resistance

4.2 Trends, Sources, and Projections

- Climate change is impacting wind speeds in complex ways with some areas experiencing increases and others decreases
- Generally, climate change is expected to alter the temperature difference between the equator and the poles potentially weakening some wind patterns
- However, other factors like natural climate cycles can also play a significant role and some areas are seeing increases in wind speeds
- As the Arctic warms faster than the tropics the temperature difference that drives winds is reduced potentially leading to slower average wind speeds in some regions
- Studies have shown that winds across parts of North America, Europe, and Asia have been increasing in recent decades
- Climate change is expected to intensify some extreme weather events including wind gusts associated with thunderstorms and hurricanes
- Climate change could affect wind energy resources with some regions experiencing decreases in wind speeds and others potentially seeing increases depending on the specific location

4.3 Factors influencing Wind Speed Changes

- The temperature difference between the equator and the poles is a major driver of wind pattern
- Changes in the temperature gradient due to climate change can alter wind speeds
- Natural climate cycles like the Pacific Decadal Oscillation (PDO) can also influence wind speed trends
- Urban development and other land-use changes can increase surface roughness which can affect wind speeds near the ground
- Warmer ocean waters can act as fuel for hurricanes leading to increased wind speeds in these storms

- A study on climate change found that the fastest jet stream winds will accelerate with warming temperatures
- Climate change is linked to increases in hurricane intensity including wind speeds due to warmer ocean temperatures

4.4 General Effects of Wind on Crops

- Wind effects crops both directly and indirectly
- Wind effects plant growth in three significant ways viz., transpiration, carbon dioxide intake, and mechanical breakage of leaves and branches
- Transpiration increases with wind speed up to a certain point beyond which neither it increases nor decreases
- Wind exerts a much greater influence on cuticular transpiration than on stomatal transpiration
- Therefore, plants with higher cuticular transpiration (hydrophytes) show an appreciable increase in the transpiration resulting from the action of the wind
- Under natural conditions the effect of wind on transpiration vary according to roughness as determined by the surface configuration
- In general, the effect is greatest for an isolated tall plant. Example: An increase in the transpiration rate of a single coconut tree in full sunlight of about 100 percent with a wind speed at five miles per hour is much more than few coconut plants grown together
- For any other crop also if the canopy cover is complete and the surface is even and smooth then the effect of wind on transpiration is usually small
- High extremes in wind speed result in the lodge of crops
- Changes in wind speed indirectly affects crop surfaces by changing the resistance, thus controlling the fluxes of heat, water vapor, carbon dioxide, and pollination

- Wind is also important to the application of pesticides, fertilizers, and irrigation water
- The lower pressures experienced as altitude increases have important consequences for high altitude plant life
- At high altitudes and low atmospheric pressures, the solubility of carbon dioxide and oxygen in water are reduced
- Therefore, some plants show stunted growth at higher altitudes as concentrations of oxygen and carbon dioxide reaches low concentrations
- Plants with strong root system and tough stems can live under increased wind speeds at low pressures in high altitude areas
- In overhead irrigation to crops the distribution of water in the field is very much dependent upon the wind speed at height of the nozzle
- Therefore, in designing the sprinkler for a given area the wind data taken at 2m or some other fixed height is needed
- At or near the surface of the ground the wind velocity is zero and increases with height above the surface
- The wind gradient above the ground depends on the wind speed as well as the surface conditions
- According to logarithmic wind profile equation the wind speed near the ground increases exponentially with height over a very smooth surface
- Over a moderately rough surface (short grass) the logarithmic wind profile holds true only above a hypothetical height Z_O, known as the roughness of the surface
- The wind structure over a tall crop is different from that over a short crop
    - Over a maize crop of 250cm, the wind profile changes abruptly at a height of about 220 cm slightly below (30cm) the canopy
    - Above that height the logarithmic relationship seems to hold good
    - Below that the wind speed is generally reduced

4.5 Relationship between Prevailing Wind Direction and Area Protected by Windbreaks

- Prevailing wind direction significantly influences the effectiveness and placement of windbreaks, directly impacting the area they protect. Windbreaks should ideally be oriented perpendicular (at right angles) to the direction of the prevailing winds for maximum protection. This alignment maximizes the protected zone, creating a "Wind shadow" on the downwind (leeward) side where wind speed is considerably reduced **(Fig: 4.1).** The other factors are

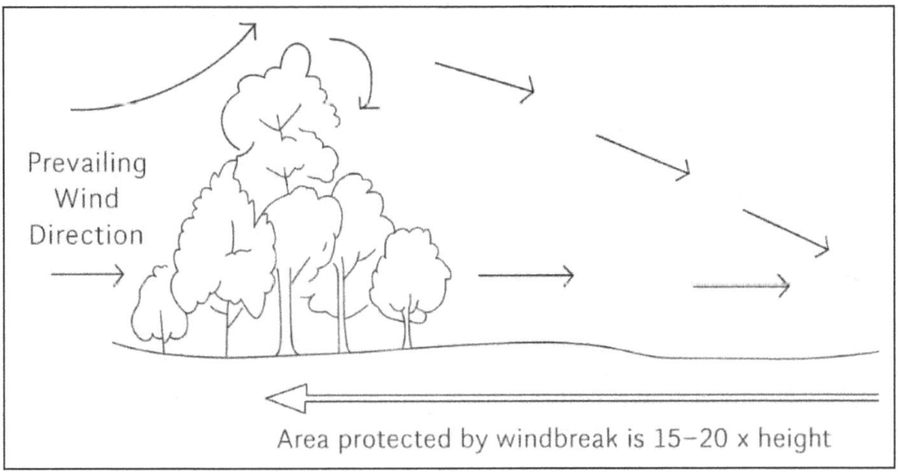

Fig. 4.1 Relationship between prevailing wind direction and area protected.
Source: Unknown. Class lecture notes collection

- Orientation: For optimal protection, a windbreak should be planted perpendicular to the prevailing winds
- Protected area: The extent of the protected area downwind (leeward) of the windbreak can reach up to 15-20 times the height of the windbreak (15-20H). The most effective protection occurs within 10 times the height (10H) of the windbreak
- Windbreak height: The height of the tallest row within the windbreak, or 'H', is the primary factor determining the size of the protected zone

- Density and porosity: Windbreak density, defined as the ratio of solid to total area, also affects protection. Optimal densities (40-60%) allow for desired airflow, reducing turbulence and extending the protected area further downwind. Very dense or solid windbreaks can create excessive turbulence directly behind them, which may reduce effectiveness further downwind
- Length and continuity: The windbreak should be at least 10 times longer than it's height to minimize the influence of wind wrapping around the ends and creating turbulence. Gaps within the windbreak can also reduce effectiveness by creating wind tunnels
- Seasonal variation: Potential seasonal shifts in prevailing wind direction have to be considered, and windbreaks have to be planned accordingly. In areas with variable winds, multiple-leg windbreaks may offer more comprehensive protection
- Understanding the local prevailing wind direction is the first step in designing an effective windbreak to achieve the desired goals, such as protecting crops, livestock, etc.

4.6 Generalized Wind Profile of a Short Crop

- In short crops, the wind profile typically exhibits a logarithmic increase in wind speed with height above the ground **(Fig: 4.2)**

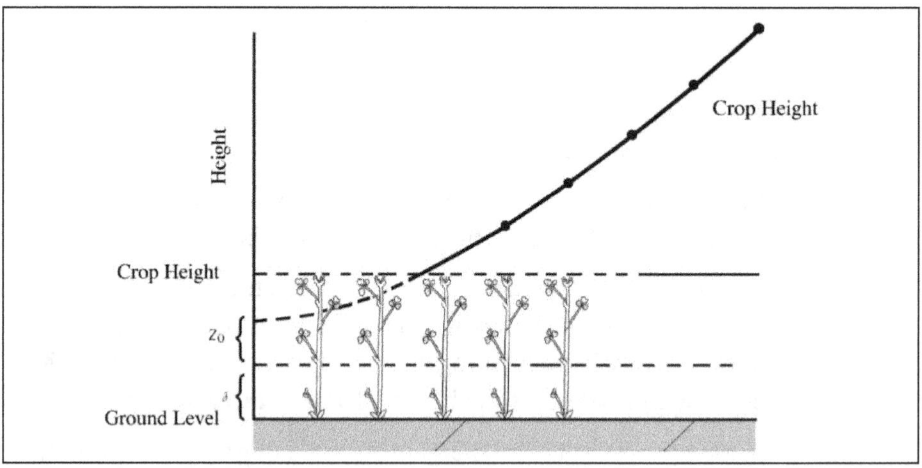

Fig. 4.2 Generalized wind profile of a short crop.
Source: Unknown. Class lecture notes collection

- However, this ideal profile is significantly influenced by the crop canopy itself
- Logarithmic profile above canopy: Mean wind speed increases roughly logarithmically with height above the top of the canopy
- Reduced wind speed within canopy: The crop canopy creates resistance, causing a significant reduction in wind speed within the canopy itself compared to the open air above
- Roughness Length (Z_O) and Zero Plane Displacement (d): These parameters describe the aerodynamic characteristics of the crop surface and influence the wind profile. Roughness length (Z_O) represents the height at which the wind speed effectively becomes zero due to surface friction. Zero-plane displacement (d) accounts for the effective height of the canopy, above which the logarithmic profile begins. Both (Z_O) and (d) vary with crop height and growth stage
- Turbulence within canopy: Airflow within the crop canopy is characterized by turbulence, which arises from mechanical friction between the wind and the vegetation. This turbulence is dominated by large-scale eddies influenced by the canopy structure
- Factors influencing wind profile
    - Crop height and density: Taller and denser crops generally lead to a greater reduction in wind speed within the canopy
    - Leaf Area Index (LAI): A higher LAI, representing the total leaf area per unit ground area, increases the canopy's resistance to wind, further influencing the wind profile
    - Wind speed and direction: Ambient wind speed and direction, as well as features like windbreaks, also affect the wind profile within and above the crop
- A short crop significantly modifies the atmospheric wind profile, creating a zone of reduced wind speed and increased turbulence within the canopy, with a logarithmic wind profile developing. Understanding this profile is critical for assessing various agricultural aspects, including evapotranspiration, pest and disease spread, and crop damage risk

4.7 Actual Wind Speeds at various Heights before and behind Barrier

- Windbreaks significantly alter wind speed both before and behind the barrier, with the most pronounced effects observed near the ground surface **(Fig: 4.3)**

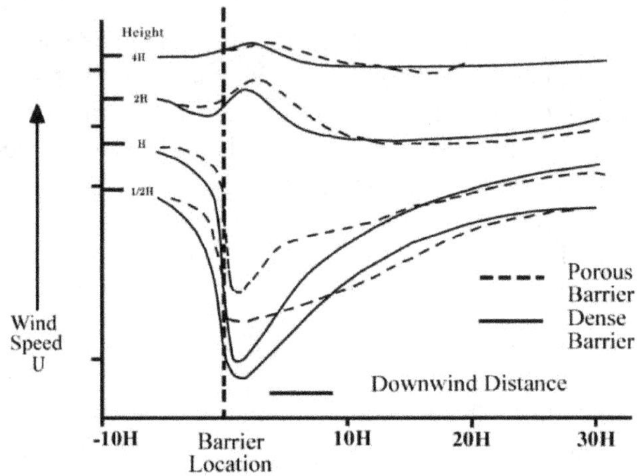

Fig. 4.3 Actual wind speeds at various heights before and behind a wind barrier.
Source: Unknown. Class lecture notes collection

- Before the barrier (windward side)
  - Wind speed reduction is measurable 2 to 4 times the height of the windbreak (H) upwind
  - The magnitude of this reduction is less significant compared to the leeward side
- Behind the barrier (leeward side)
  - Significant wind speed reduction occurs up to 30 times the windbreak's height (30H) downwind. For example, a 30-foot tall windbreak can reduce wind speed for up to 900 feet on the leeward side
- Height (H): Taller windbreaks provide a larger protected area

- Porus barriers and dense barriers: These differ on the extent and magnitude of wind speed reduction
    - Porosity: Refers to the amount of open space within the barrier. Lower porosity often results in higher/ maximum wind reductions but create more turbulence
    - Density: Windbreak density (ratio of solid area to total area) plays a crucial role
        - Optimal dense (40-60%) provides the largest protected area with good wind speed reduction
        - Very dense windbreaks (60-80%) lead to greater reduction but a shorter protected zone due to turbulence
        - Low density (<20%) offers minimal wind speed reduction
- Orientation: Windbreaks are most effective when oriented perpendicular to prevailing winds
- Length: Longer windbreaks (at least 10 times the height) reduce the impact of end turbulence
- Continuity: Gaps within a windbreak act as funnels, increasing wind speed locally
- Wind speed at different heights
    - Windbreaks primarily reduce wind speeds at lower heights, particularly near the ground surface
    - Wind speeds generally increase with height above the ground surface, even in sheltered areas

4.8 Influence of Solid Barrier on Airflow in Windbreaks

- When wind encounters a solid barrier (like a densely packed windbreak or a solid fence) it largely flows up and over it **(Fig. 4.4)**

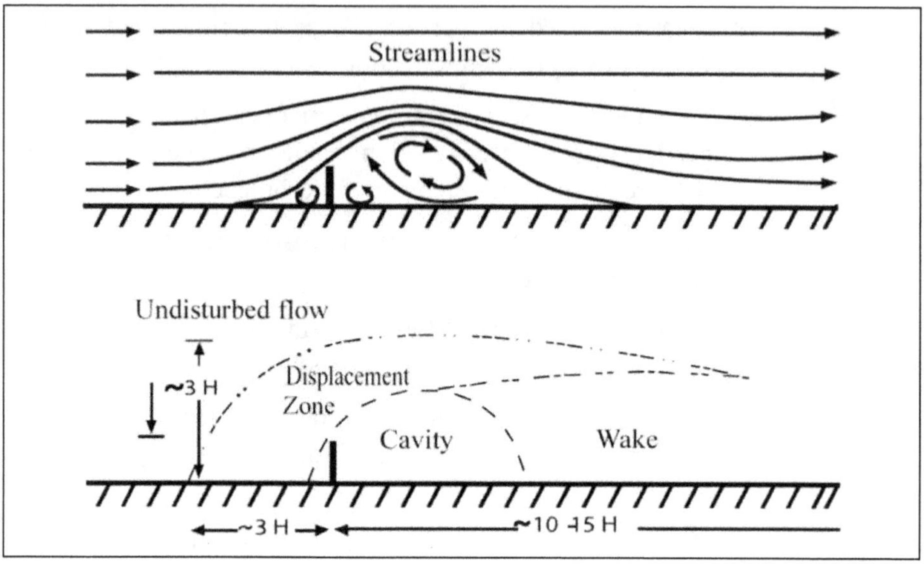

Fig. 4.4 Influence of solid barrier on airflow in windbreaks.
Source: Unknown. Class lecture notes collection

- Pressure buildup: Air pressure increases on the windward side
- Reduced downwind protection: A low-pressure zone (cavity) on the leeward (downwind) side can create turbulence, pulling air down and limiting the extent of wind speed reduction
- Turbulence and recirculation: Very solid barriers can lead to significant turbulence downwind, potentially causing airflow recirculation. This might even harm sensitive plants
- Limited protected area: While providing a significant reduction in wind speed near the barrier, the overall protected area behind a solid barrier might be shorter (10-15H) compared to more permeable windbreaks
- Snow distribution: Solid windbreaks can cause deep, concentrated snowdrifts near the barrier (typically within 3 to 5 times the height), rather than distributing snow evenly over a larger area
- A solid barrier maximizes wind reduction immediately behind it

- The displacement zone starts at 3H windward side and merger with "Wake" zone around 15H
- A solid barrier has the potential for turbulence and a more limited protected area
- Therefore, often moderately porous windbreaks (around 40-60 %) are more effective for purposes like agricultural protection or snow distribution

4.9 Basic Micrometeorological Terminology and Laws of Wind for Climate Change Studies

- Momentum transfer within the canopies
  - o Wind profile and Renold's stress
    - Basic characteristic features of wind profile are "S-shaped" in all the crop canopies, forests, etc. The resultant exponential flux is also "S-shaped". These are fundamental profiles, the relationships of which are well established with momentum equations. The nature of momentum transfer within the canopies can well be understood by momentum equations
    - The exponential Renold's stress profiles can successfully be predicted by LAI profile alone
    - The characteristics of the "S-shaped" wind profile can be theoretically explained by the plant morphology and local drag co-efficient distribution. Maximum drag co-efficient are usually located around maximum LAI
    - A universal relationship of Renold's stress between the top and bottom of the canopy exist for all crops
    - This universal relationship can be used to understand what percentage of the Renold's stress at the top of the canopy is absorbed by the whole canopy layer from the observed LAI values alone

- There should always be a consistence that satisfy the continuity requirements of Renold's stress, mean wind speed, and local drag co-efficient at the top of the canopy

- Momentum flux – Newton's Law of viscosity
    - The resistance of a fluid to flow is known as "Viscosity", informally "Thickness"
    - Newton's law of viscosity describes how a fluid's internal friction, or viscosity, relates to the shear stress and shear rate when the fluid is in motion
    - In simpler terms, Newton's law of viscosity states that "The force needed to make a fluid flow (shear stress) is directly proportional to (shear rate) i.e., how quickly the fluid is deforming
    - This relationship is constant for a given fluid at a specific temperature and pressure, and it is this constant that defines the fluid's viscosity
    - Newton's Law of Viscosity helps to understand how fluids resist flowing or how "Thick" or "Thin" they are
    - Fundamentals of Newton's Law
        - Imagine a fluid (water or honey) flowing in layers, one on top of the other
        - When a force is applied to make the fluid flow, there is a kind of friction between these layers, resisting their movement past each other. This resistance is called "Viscosity"
        - As stated above, Newton's Law of Viscosity states that resistance, or shear stress, is directly proportional to velocity gradient i.e., how quickly the layers are sliding past each other Stress =
        - Viscosity × Rate of flow
        - Viscosity = Stress × Time
        - All fluids have viscosity

- Fluids are made of many molecules
- The more the resistance the higher the viscosity
- A fluid that is highly viscous has a high resistance
- Water is much less in viscosity than oil
- Many fluids have a very strong temperature dependent viscosity. Example: On boiling oil its viscosity decreases
- Fluids put a stress force that prevents the flow
- Water flows freely and quickly; honey flows slower than water. Inference: The easier a fluid moves the lower the viscosity
- In high viscous fluids strong forces of attraction exist. When heated, attractive forces becomes less
- Different magnitudes of relative movement. Relative movement = viscosity. More viscosity = Strong forces of attraction. Less viscosity = Weak forces of attraction. Example: A boat moves faster on water than oil. Viscal force reduces movement
- Pressure and temperature are also considered to understand the viscosity of a fluid
- If viscosity does not change with pressure, it is called as "Newtonian fluid"
- If viscosity changes as temperature changes it is known as "Non-Newtonian fluid"
- Viscosity is also related to density of a fluid. Density describes the space "Volume" that is occupied by the particles of a substance and is measured in kg/m^3
- Viscosity is measured in pascal seconds a measure of force equalling one kilogram per meter per second

- Richardson number
    - This is a dimensionless number and is a measure of the importance of buoyance forces (thermal convection) in producing turbulence as compared with frictional forces
    - The Richardson number equals zero under neutral conditions
    - It is negative under super adiabatic conditions and positive during temperature inversion
- Roughness (Z_O) and Zero plane displacement (d)
    - Both roughness (Z_O) and Zero plane displacement (d) are geometric constants of the surface
    - Under neutral conditions, the relationship between the wind velocity and logarithm of height is linear, and the intercept is the "Roughness (Z_O)"
    - However, over a tall crop, the relationship is curvilinear, and the departure from straight line is known as "Zero plane displacement (d)"
    - Under both lapse and inversion conditions the relationships between wind speed and the logarithm of height are also curvilinear
    - For relatively smooth surfaces the roughness parameter is relatively constant over a range of wind speeds
    - Typical values of roughness are as follows: Ice 0 cm; Open water 0.02 to 0.06 cm; Snow surface 0.1 to 0.6 cm; Short grass 0.6 to 4.0 cm; Long grass 4.0 to 10.0 cm
    - For tall crops, which bend and weave with wind, both Z_O and d are subjected to considerable variations. Example: For a rice crop of 90 cm in height, the value of d varies from 35 to 90 cms and Z_O from 7 to 18 cms for wind speeds upto 10cm sec^{-1}
    - This complicated relationship is the result of the bending of the stalks and the thickening of the canopy which causes a variation in the drag co-efficient
    - The value of "d" increases with wind speed and wind direction

- The value of "d" is a function of the height as well as the density and mechanical properties of the plant
- Plants that stand erect in the wind should have a higher value of "d"
- In general, the value of "d" ranges from 50 to 80 percent of plant height depending upon wind speed, wind direction, and mechanical properties of plants

- Significance of roughness
  - An increase in roughness causes lowering of maximum temperature during the daytime and a rise in minimum temperature at night
  - The rougher the surface the greater the mixing and swirling
  - Therefore, the evapo-transpiration of a rough surface exceed that of a smooth surface especially in areas of string advection

4.10 Wind Techniques for Climate Change Mitigation and Adaptation

- Wind Breaks
  - Wind breaks in modification of microclimate
    - Humans/farmers have modified the microclimate of fields for thousands of years by irrigation
    - They have also used wind breaks for many years
    - The major rationale for using wind breaks is to reduce winds resulting in less soil erosion and direct damage to crop yields
    - A major cause of crop loss in some windy regions is lodge or the blowing over of top-heavy crops prior to harvest
    - Wind breaks have also been used to keep animals safe from extremes in cold and hot weather

  - Other effects that have been reported include
    - Decrease of evapotranspiration of crops
    - Increase in photosynthesis and yield of crops
- Effects of wind breaks on wind speeds
  - The primary purpose of wind breaks is to reduce wind speeds within a field
  - Reductions in speed of up to 75% may be obtained by using a wind break
  - The reduction effect is greatest at low levels at heights below that of the wind break
- Other benefits of wind breaks
  - There is overwhelming evidence to support positive benefits of wind breaks
  - Generally, most benefits occur within 10 h on the leeward side (Maximum 3-6 h) or within 0-3 h on the windward side
  - The improvement in crop quality is a major benefit, ranging from specific benefits such as an increase in sugar content of sugar beet and optimum nicotine content of tobacco
  - There is a general reduction in physical abrasion and earlier ripening of most crops
  - Average yield increase ranges from 5 to 50 per cent
  - Vegetables and specialty crops are highly susceptible to wind abrasion
  - In most cases crop quality is directly related to the degree of protection
  - The crop quality determines the price for most vegetable and specialty crops
  - Therefore, extensive wind protection systems are justified
  - In orchards enhanced microclimate improves conditions for pollination and fruit set which in turn results in greater yields

- Reduction in wind speed within the orchard reduces the amount of mechanical damage caused by whipping of leaves, branches, buds, flowers, and fruits
- In addition, wind breaks reduce tree deformation and root breakage in young fruit trees
- Winter wheat, barley, rye, millet, and alfalfa are highly responsive to protection
- Windbreak height, longevity, field width, and windbreak orientation are major considerations in determining the effect of windbreaks on crop yields

o Considerations

- Land use: A portion of the land is dedicated to the windbreak, reducing the area available for crops
- Competition: Trees within the windbreak can compete with adjacent crops for water and nutrients, particularly in areas close to the barrier
- Shading: Windbreaks can create shade, potentially hindering growth for crops in their immediate vicinity
- Pests: In some cases, windbreaks might provide habitat for certain pest species

- However, research suggests that the benefits of well-designed and managed windbreaks generally outweigh these drawbacks, resulting in a net increase in overall crop productivity and quality over time, especially in dry or windy conditions

4.11 Addressing Climate Change Challenges on Wind

- The individuals, families, communities, governments, organisations etc., have the best sciences available to plan for and implement climate-smart and resilient practices for negative impacts of climate change driven wind issues. Two critically important approaches which are available to managing climate resilience are mitigation and adaptation. These strategies are inter-related and must be pursued simultaneously as both can potentially improve resilience to the changing climate.

Chapter 5

Climate Change and Temperature

"According to Vedic thought, the concept of "Temperature" transcends a simple physical measurement. Instead, it is understood as the dynamic interplay of heat (agni) and cold signifying cosmic energy, transformation, and the fundamental duality of existence. This cosmic transformations of contrasting energies shape both the universe and the human experience impacting physical processes and influencing the journey towards spiritual wisdom"

5.1 Introduction to Air Temperature

- The temperature of the Earth's surface is primarily controlled by the influence of the Sun and the properties of the atmosphere
- The earth is a spherical body at a large distance from the Sun
- The Sun has different apparent elevations when viewed from the Earth's surface at different latitudes
- The elevation of the Sun and the time of the year act in combination to determine the temperature of the Earth
- The effective average temperature of the Earth Atmosphere System (EAS) without the atmosphere is 253 degrees Kelvin
- The average temperature of the Earth is 283degrees Kelvin

- That means, the presence of atmosphere keeps the earth's surface temperature about 30degrees Kelvin above what would be without the atmosphere
- The Sun is considered as a black body which absorbs and emits radiation efficiently. The peak wavelength of emitted radiation (0.48 microns) is related to its absolute temperature (6000 degrees Kelvin) as described by Wein's Law
- The same law applies to terrestrial objects and so an object at 293 K (20 °C) will have a peak emission at 9.89 microns. As mentioned in the solar radiation chapter this is one of the reasons for declaring 20 °C as the most optimum temperature for crop plants for their growth and development
- Large quantities of energy are transferred between the Earth surface and the air by the process of convection and the transfer (Flow) determines air temperature
- The mechanism of transfer is molecular diffusion rather than direct conduction
- The daily oscillation of air temperature describes a sine curve, with the minimum normally occurring in the early morning hours around sunrise and the maximum occurring sometime after peaking solar and net radiation
- The patterns of air temperature waves are not necessarily regular when examined based on individual days, especially in climates that feature frequent frontal passages, irregular cloudiness, or strong advection of sensible heat generated in other regions
- The afternoon temperature lag is a result of the balance between incoming and outgoing radiation
- From sunrise on, a considerable amount of radiant energy is required to heat the soil and crop which are cool at that time
- Until these surfaces become warm relative to the air above, no net sensible heat flux to air occurs
- The reason for annual lag is no different than for the daily lag
- As the portion of energy flux decreases and the soil becomes warm relative to the environment, more energy is converted to sensible heat

- During spring and early summer, a large portion of the incident solar energy flows into the soil which reaches its lowest temperature by the end of winter
- A terrestrial body always absorbs and emits longwave radiation even when solar radiation is absent
- Longwave radiation transfer is important in determining the thermal or temperature relationships of crop plants in their environment
- The water vapor, carbon dioxide, and some of the cholo-fluro methanes and other materials of organic origin are essentially transparent to shortwave radiation, but are effective absorbers in the longwave radiation (thermal infrared region) of the electromagnetic spectrum
- Hence, any increase in the concentration of these substances in the atmosphere is expected to increase global temperature through a trapping of the terrestrially emitted longwave radiation

5.2 Trends, Sources, and Projections

- Instrumental observations over the past 157 years show that temperatures at the surface have risen globally with important regional variations
- Global warming in the last century occurred in two phases, from the 1910s to the 1940s (0.35°C) and more strongly from the 1970s to 2005 (0.55°C)
- Global observations above the surface since the late 1950s show that the troposphere (About 10 km) has warmed at a slightly greater rate than the surface while the stratosphere (About 10-30 km) has cooled markedly since 1979
- This is in accord with physical science principles and results of most models
- The temperature in the first two decades of the 21st century (2001-2020) was 0.99 °C higher than 1850-1900

- Global surface temperature was 1.09 °C higher in 2011– 2020 than 1850–1900 with larger increases over land 1.59 °C than over the oceans 0.88 °C
- Global surface temperature has increased faster since 1970 than in any other 50-year period over at least the last 2000 years
- Temperatures during the most recent decade (2011–2020) exceeded those of the most recent multi-century warm period, around 6500 years ago 0.2°C to 1°C relative to 1850– 1900
- The temperature increase will cross the 1.5 degree centigrade threshold by 2040, which was projected much later
- The increase in air temperature varies according to the likely prevalent scenario which includes factors like population density, rate of economic growth, and ratio of use of fossil and non-fossil energy sources
- another interesting feature is that the rise in minimum temperature will be more than that of maximum leading to a steady decrease in the daily temperature range over the years
- This non-symmetry in increase of minimum and maximum temperatures has great agronomic and crop physiological significance
- Continued greenhouse gas emissions leads to increasing global warming with the best estimate of reaching 1.5 degrees centigrade in the near term (2021- 2040) and very likely to exceed 1.5 degrees centigrade
- This is mainly due to increased cumulative carbon dioxide emissions
- With every additional amount of global warming, changes get larger (at 1.5, at 2.0, at 4.0 degrees centigrade)

5.2.1 Latest IPCC Reports on Temperature

- The Intergovernmental Panel on Climate Change (IPCC) reports consistently highlight a clear and undeniable trend of global temperature increase, attributing it unequivocally to human activities, primarily the emission of greenhouse gases

- Global surface temperature changes relative to 1850-1900, is projected by IPCC on all emission scenarios viz., very high, high, intermediate, low, and very low. **(Fig:5.1)**

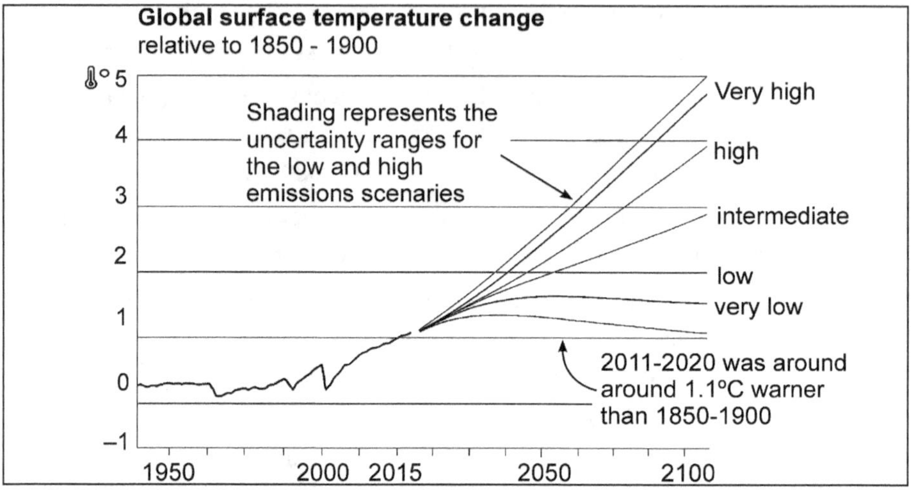

Fig. 5.1 Global surface temperature changes relative to 1850-1900.
Source: IPCC 2021, 2022 and 2023

- Observed warming: The Earth has warmed by approximately 1.1°C (2°F) since 1850-1900, with a significant increase in the rate of warming observed since 1982
- Unprecedented changes: This warming has driven changes to the Earth's climate system that are unprecedented in recent human history, affecting every region of the world
- Regional variations: While the global average temperature is increasing, warming is not uniform across the planet. Land areas are warming faster than oceans, and the Arctic is experiencing warming at a rate more than twice the global average
- Increased extreme events: This warming has led to more frequent and intense heat waves, heavy rainfall events, and droughts in many regions
- Future projections: Even with ambitious emission reductions, global warming is projected to continue, with the average global temperature expected to reach or exceed 1.5°C in the coming decades, potentially sooner under high emissions scenarios

- Impact of warming: Every increment of warming is expected to bring increased risks and impacts and threats to ecosystems and biodiversity
- Irreversible changes: Some changes, like sea level rise, are already set in motion and are irreversible over hundreds to thousands of years
- Urgency of action: Limiting global warming to 1.5°C will require immediate, rapid, and large-scale reductions in greenhouse gas emissions

5.2.2 The Ten (10) Hottest Years on Record

- The years 2015-2024 are the ten warmest years on record globally, according to NOAA and the World Meteorological Organization (WMO)
- NOAA data shows Earth's average land and occan surface temperature in 2024 was 2.32 degrees F (1.29 degrees C) above the 20th-century average, marking the highest global temperature recorded between 1850 and 2024
- Earlier reports of NASA GISS &NOAA NCEI global temperature anomalies (°C) averaged and adjusted to early industrial base line (1881-1910) indicate that 10 warmest years occurred between 1998 and 2017 **(Fig:5.2).**

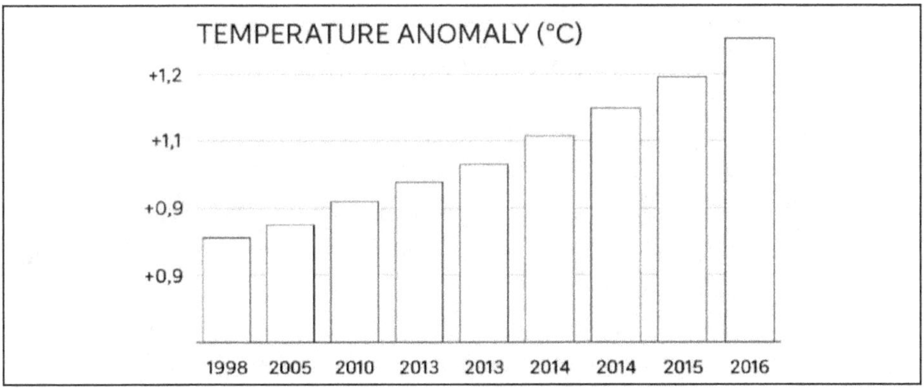

Fig. 5.2 Ten(10) hottest years globally.
Source: NASA GISS &NOAA NCEI global temperature anomalies (°C) averaged and adjusted to early industrial base line (1881-1910)

- These reports highlight a clear and sustained trend of rising global temperatures over the past decade, emphasizing the urgent need for action on climate change

5.3 General Effects of Air Temperature on Crops

5.3.1 Generalised Crop Growth Response to Temperature

- Crops exhibit a generalized, but diverse, response to temperature, influencing growth and yield **(Fig: 5.3)**

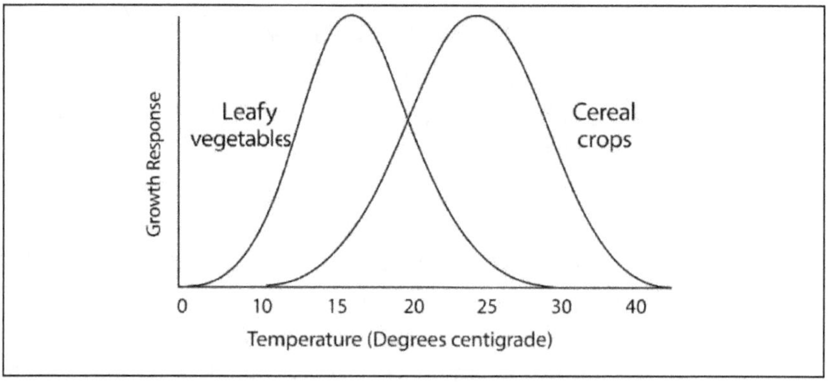

Fig. 5.3 Generalised crop growth response to temperature.
Source: Unknown. Class lecture notes collection

- The generalized optimum temperature for leafy vegetables is 15 to 20 degrees centigrade and for cereals it is 20 to 30 degrees centigrade
- The optimal range generally accelerates metabolic processes leading to faster development and maturation
- Above optimal temperatures result in heat stress. It,
  - Reduces photosynthesis due to enzyme denaturation and damage to chloroplasts
  - Increases respiration, leading to carbon loss and decreased biomass accumulation, particularly at night
  - Shortens the reproductive phases (flowering and grain filling) limiting the time for nutrient accumulation and ultimately reducing yield
  - Can lead to water stress and stomatal closure, further impeding carbon dioxide uptake and cooling

  - o Impacts include flower abortion, poor pollen viability, reduced fruit set, and decreased seed weight and quality

- Currently temperatures in India are one degree centigrade above optimal for rice and wheat
- Therefore, further global warming adversely affects rice and wheat crop yields necessitating technological interventions
- Understanding the specific temperature requirements and stress thresholds of different crops is crucial for optimizing agricultural practices and adapting to changing climate conditions

5.3.2 Growing Degree Unit Assumed Relationship between Plant Growth and Temperature

- Air temperature plays a critical role in determining the length of each phenological phase of crops
- Each growth phase of a crop must meet its heat unit requirements before transitioning to the next phase
- A heat unit is also known as a growing degree day (GDD), is a measure used in agriculture to estimate crop development based on temperature accumulation **(Fig: 5.4)**

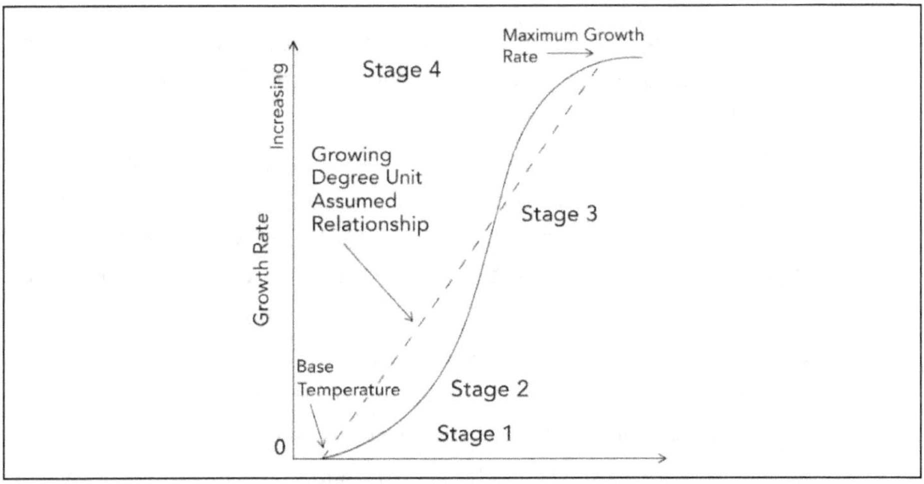

Fig. 5.4 Growing degree unit assumed relationship between plant growth and temperature.

Source: Unknown. Class lecture notes collection

- As temperature increases the growth rate increases from stage 1 to stage 4
- The maximum growth rate occurs at stage 4
- The GDD assumed relationship follows a linear progression until stage 4
- Crop duration and durations of various phenological phases of a crop variety are determined by its accumulation of growing degree days over a specified base temperature
- Each plant species has a minimum temperature threshold, known as the base temperature, below which little or no growth and development occurs
- GDD is calculated by subtracting the base temperature for a specific crop from the average daily temperature
- If the average temperature is below the base temperature, the GDD for that day is zero
- By summing the daily GDD values over the growing season, a cumulative GDD is obtained
- The base temperature for accumulation of day degrees varies amongst crops
- For a given cultivar and a given temperature increase, the extent of reduction in crop duration varies with the level of ambient air temperatures
- Thus, a given rise in temperature affects reduction in crop duration of (a) different crops differently in the same location (b) of the same crop differently in different locations in the same season and (c) of same crop in different seasons in the same location
- It is this feature that necessitates assessment of the impact of climate change on crop prospects to be carried out on a crop-wise, location-wise and season-wise basis
- Applications of GDD: The accumulated heat unit measurement allows farmers and scientists to
    - Predict crop growth stages and maturity: Helping to schedule planting, irrigation, pest management, and harvest times

- o Compare crop development across locations and seasons: Help to assess which areas are best suited for certain crops
- o Estimate potential crop yields

- Limitations: While GDD is a valuable tool, other factors like soil moisture availability and light can also impact growth and development. Additionally, some GDD models incorporate upper temperature thresholds, as excessively high temperatures can also hinder crop development

5.3.3 Generalized Relationship between Development Time and Temperature of Crops

- As detailed above, the crop development time is directly tied to the accumulation of heat units, often quantified as Growing Degree Days (GDD), and is thus strongly influenced by temperature **(Fig:5.5)**

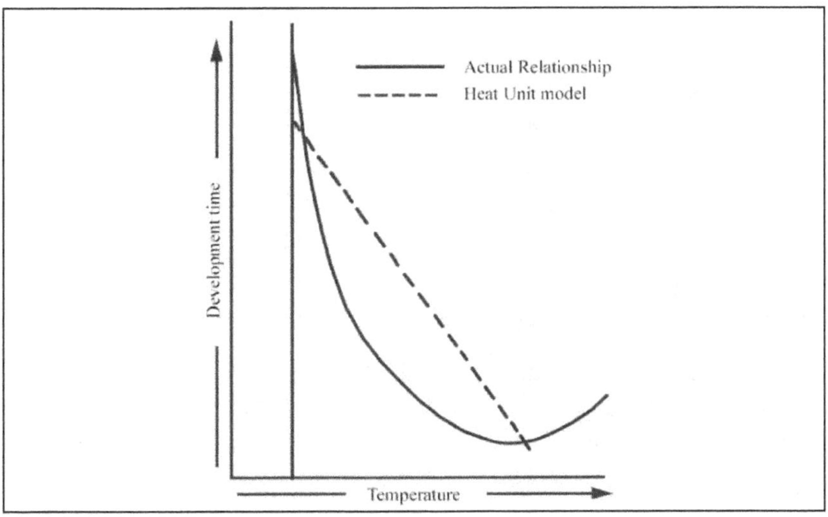

Fig. 5.5 Generalized relationship between development time and temperature of crops.

Source: Unknown. Class lecture notes collection

- The actual relationship between temperature and development time indicates that as the temperature increases above the optimum, the development time decreases

- The heat unit model also indicates the same
- Essentially, warmer temperatures (within the optimum range) generally lead to faster development, while cooler temperatures slow it down
- This relationship is crucial for agricultural planning and understanding how crops respond to changing climates

5.3.4 Generalized Effect of Temperature on RGR in Tropical, Temperate, and Sub Polar Regions

- In the context of agriculture, relative crop growth rate (RGR) varies across climatic regions due to differing environmental conditions and crop adaptations **(Fig:5.6)**

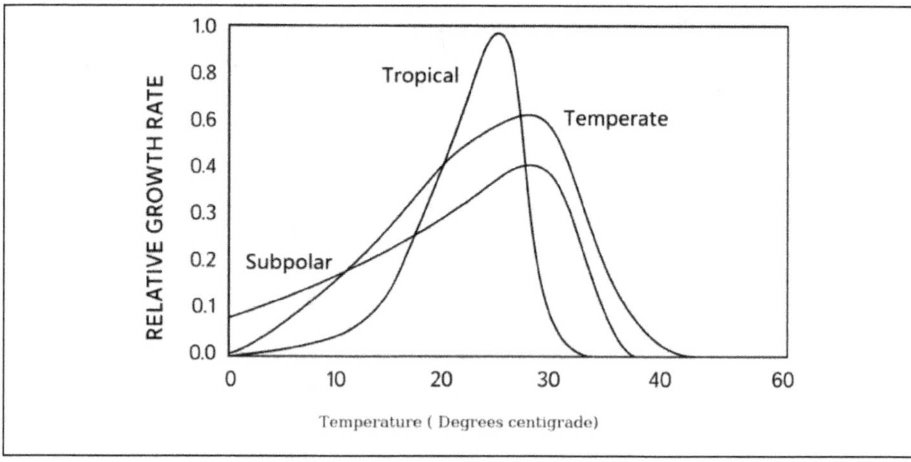

Fig. 5.6 Generalized relationship of plant relative growth rate and temperatures in tropical, temperate and subpolar regions.

Source: Unknown. Class lecture notes collection

- Tropical regions
  - o Generally high RGR: Tropical regions, characterized by high temperatures and abundant rainfall (in humid areas), support crops with high growth rates year-round
  - o Optimal conditions: Favorable temperatures, consistent moisture, and ample sunlight facilitate rapid photosynthesis and biomass accumulation in many tropical crops

- Challenges: However, excessive heat can reduce yields, and uneven rainfall distribution necessitates irrigation in drier areas

- Temperate regions
  - Seasonal variability: Temperate zones experience distinct seasons, with RGR significantly influenced by temperature and day length variations throughout the growing season. Supports crops medium growth rates
  - Potential for increases: Moderate warming in temperate zones, where temperatures might be below the optimal for some crops, can potentially increase average yields
  - Adaptation strategies: Farmers utilize practices like selecting suitable crop varieties with appropriate thermal time and vernalization requirements, adjusting planting dates, and employing irrigation to manage growth and yield in response to changing climate
- Subpolar regions
  - Limited growing season: Subpolar regions face a significantly shorter growing season due to low temperatures and long periods of snow cover. Low warming supports the crops with less growth rates than temperate regions much more less growth rates than tropical regions
  - Adaptation for rapid growth: Plants adapted to these regions often exhibit adaptations that allow them to maximize growth during the brief periods of favorable conditions
  - Vulnerability to extremes: Crops in these areas are highly sensitive to sudden cold snaps and shorter growth periods caused by climate shifts
  - Relative crop growth rates are intricately linked to the climatic conditions of each region
  - Tropical regions typically exhibit high RGRs due to warmth and moisture, while temperate zones experience seasonal fluctuations and subpolar areas present a challenging environment requiring specific adaptations for survival and growth

5.3.5 Temperature and Photorespiration

- When the temperature increases to three (3) degrees centigrade above normal, the beneficial effect of doubled CO_2 gets canceled out for any crop in any season and any location
- A doubling of CO_2 leads to an increase of temperature from 3 to 5 degrees centigrade. Thus, in real-time an increase in elevated CO_2 leads to a decline in crop yields
- In global warming, minimum temperatures are increasing more than the maximum temperatures
- Increase in minimum temperatures lead to increased respiratory depletion of photosynthates
- The maintenance respiration for rice is 7 per cent of photosynthesis and about 10 per cent of photosynthesis for other crops
- Thus, increased temperature reduces biomass production in crops
- Increased temperatures increases the photorespiration of C_3 crops
- Thus, the effects of increased temperature on yield reduction will be more for C_3 than C_4 crops

5.3.6 Temperature and Added Fertilizers

- Increased temperatures lead to a decrease in response of crops to added fertilizers
- Higher air temperatures lead to higher soil temperatures which increase mineralization and gaseous loss of nitrogen
- Thus, global warming necessitates increased applications of nitrogenous fertilizers
- This in turn leads to a greater infusion of nitrous oxide into the atmosphere

5.3.7 Temperature and Frost

- Climate change driven by the increase in minimum temperatures results in fewer frost events

- Initially the increase in temperatures leads to an increase in frost-free period and earlier maturity and harvest of crops in temperate zones and benefit agricultural and forage crops
- There will be reduced and/or less severe incidence of frosts in all crop-frost prone areas
- Increasing daytime temperature leads to an increase in yields in Northern latitudes (increasing growing season length) but decreased yields in middle latitudes (due to earlier ripening)
- Increased cold stress reduces livestock eating and grazing activity
- This in turn causes reductions in yield and fertility of livestock
- In areas with cold stress, wind amplifies the problem for young animals
- Globally, most crop plants are confined to a narrower range of about 0-40 °C of air temperature
- Temperatures below O °C physically limit life by the likelihood of ice formation in biological tissues all of which have a high-water content
- Ice formation and subsequent melting disrupt cell structure
- In addition, freezing leads to the desiccation and subsequent death of the plant cells

5.3.8 Temperature and Snow Melt

- As the snow reflects 95 per cent of the incident solar radiation, melting of snow occurs with a rise in ambient air temperature
- Thus, global warming leads to melting and a reduction in the extent of snow cover
- At present glaciers are reported to be melting and shrinking
- Thus, in global warming irrigation systems dependent on water from snowmelt there will be an increase in quantum of water in the irrigation system initially
- The associated increase in thickness of freshwater layer in the oceans in deltaic areas leads to an increase in rainfall in coastal areas

- This gives a false sense of increased water availability and water security
- However, with the progress of time coastal rainfall and water availability begin to decrease at an alarming rate leading to reduction of crop production at an increasing rate in such systems

5.3.9 Cardinal Temperatures

- Regardless of favorable light and moisture conditions, plant growth ceases when temperature drops below a certain minimum value or exceeds a certain maximum value
- Between these limits there is an optimum temperature at which growth proceeds with greatest rapidity
- These three points are known as "Cardinal temperatures"
- Vant Hoff's law: Between minimum and optimum temperatures, dry matter production follows Vant Hoff law i.e., for every 10 °C rise in temperature, the rate of dry matter production doubles

Table 5.1 The cardinal temperatures for different crops (Degrees centigrade).

S.No	Crops	Minimum	Optimum	Maximum
1	Cool seasons crops (Wheat, Oats)	0-5	25-31	31-37
2	Hot season crops (Sorghum, millets)	15-18	31-37	44-50

5.3.10 Day Time and Night Time Temperatures and Crops

- As mentioned above according to the growing degree day concept "The growth and development of a plant is dependent upon the total amount of heat to which it is subjected during the lifetime, expressed as Growing Degree-Days"
  - Temperature during day is known as "Photo temperature"
  - Temperature during night is known as "Nycto temperature"
- Constant temperature during day and night compared with a high photo temperature and low nycto temperature results in considerably less growth in tomato crop

- The quality of fruits and seeds is often affected by thermo periodicity
- The sucrose concentration of both sugar beets and sugarcane increases with the decrease of night temperature
- Knowledge of ideal combination of daytime and nighttime temperatures for the growth of a plant is essential for its successful introduction and selection of proper growing season
- Whenever the microclimate coincides with the temperature requirements of plants it grows best at that time
- A humid tropical climate is characterized by its small temperature range both diurnally and annually
- Many cool weather crops desiccate within a relatively short time in the tropics, simply because they cannot adapt to the thermal regime
- High temperatures in tropics usually result in an increase in shoot-root ratio and are unfavorable for the ripening of many annual crops
- Even with the same total dry matter production, the grain yield of a rice crop is usually higher in temperate climate because of the autumn temperature. With low temperature, the transportation of carbohydrates to the grain is slow but lasts for a longer period. The senescence of leaves occurs late, and the ripening period of the grain is prolonged. This results in a higher grain weight in temperate climates than in high temperature conditions in tropical climates

5.3.11 Temperature and Flavor of Strawberries

- When strawberry plants are grown in warm or moderate temperatures the fruits are red, slightly acidic and have no strawberry flavor
- To develop flavor they must ripen at daytime temperatures of about 50 °F
- The plants must be exposed to right light and temperature conditions for at least a week during fruit set to acquire full strawberry aroma

- The experimental results indicate that the first strawberries of the season taste best
- They ripen while the early morning temperature is about 50 °F
- Later in the spring and during the summer the strawberry crop is practically without aroma because the ripening fruit does not receive the proper temperature at any time during the day
- At high latitudes, or far North, low morning temperatures occur even during summer and strawberries from Alaska and Sweden taste marvelous at any time

5.3.12 Effect of Night Temperature on Tomato Plant

- Optimal Night Temperature: For tomato plants, ideal night temperatures range roughly from 18.5 to 24.5 °C for optimal growth of roots, stems, intact plant, and sugar translocation thereby fruit set. Night temperatures slightly lower than daytime temperatures are generally preferred for optimal growth **(Fig: 5.7)**

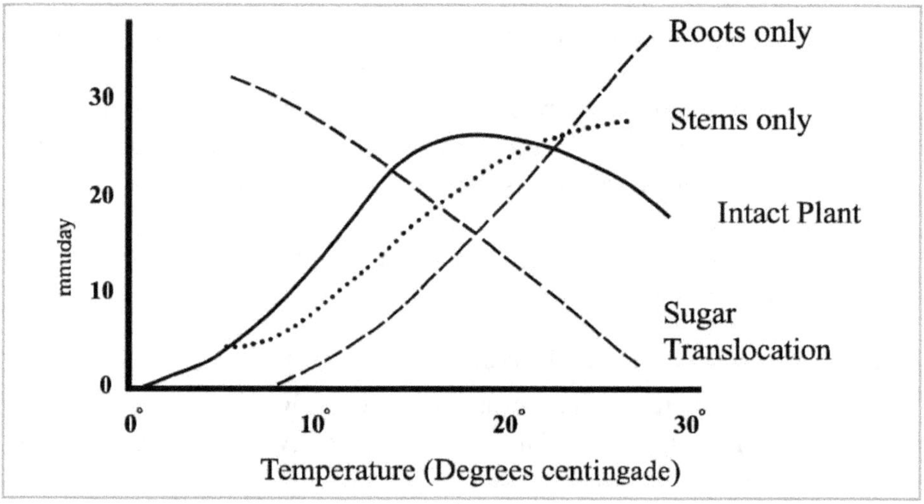

Fig. 5.7 Effect of night temperature on tomato plant.

Source: Unknown. Class lecture notes collection

- Low Night Temperatures (< 10 °C or 50 °F)
  - Reduced fruit set and abnormal flowers: Temperatures below 10 °C during flowering significantly hinder growth of roots, stems, intact plant, and sugar translocation thereby fruit set and leads to abnormal flower development
  - Chilling injury: Extended periods (weeks) below 15.5 °C (60 °F) affect growth of roots, stems, intact plant, and sugar translocation thereby flowering with reduced fruit set and potentially misshapen or "Cat faced" fruit. Chilling injury also causes stunted growth, wilting, leaf damage, and increased disease susceptibility. Seedlings are more sensitive to chilling injury below 60°F
  - Stunted growth and reduced photosynthesis: Consistently low night temperatures result in stunted growth and reduce the efficiency of photosynthesis
- High Night Temperatures (> 26.5 °C or 80 °F)
  - Reduced fruit set: Night temperatures above 26.5 °C (80°F) decrease growth of roots, stems, intact plant, and sugar translocation thereby potentially lower the number of flowers produced and subsequent fruit set
  - Increased respiration: High night temperatures lead to increased respiration rates, potentially consuming stored carbohydrates and negatively impacting growth and yield
- Both excessively high and low night temperatures stress the plant, negatively affecting optimal growth of roots, stems, intact plant, and sugar translocation thereby flowering, pollination, fruit set, and overall yield

5.4 Leaf Temperature and Crop Growth

- The radiation load on the leaf is disposed off in 3 ways (a) transpiration, (b) heat loss to the air, and (c) heat storage in the leaf
- The temperature of a leaf exposed to sunlight is often well above that of the air, especially for a thick leaf that is not transpiring in still air

- Under conditions of high insolation and high humidity leaves commonly have temperatures 15 °C above air temperature
- Leaf temperatures as high as 52 °C, some 10 °C above air temperature occur under exceptional conditions during summer months in case of perennial plantation crops
- For plants that are exposed to the Sun but do not suffer from water stress, the difference between air and leaf temperatures is usually small
- The tendency for equality (between leaves and air) is 33 °C. Below 33 °C the leaves tend to be warmer than the air. Above 33 °C the air temperature is more than the leaf temperature
- The temperature of a sunlit leaf responds quickly to the changes in wind speed and cloud cover
- If the temperature of the leaf is 10 °C above air temperature with no breeze, then if a breeze of 3kmph occurs the temperature of leaf drops by 6 °C and maintains 4 °C
- If leaf temperature is 48 °C, air temperature is 28 °C, if a dense cloud comes across the Sun and the leaf temperature will be 28 °C i.e., air and leaf temperature become equal in 30 seconds
- In many instances, the temperature of upper and lower leaves is ±0.2 °C
- In some thick leaves the upper sunlit surface to be as much as 2 °C higher than the temperature of lower surface. This difference is due to the stomatal transpiration taking place primarily from the lower surface of the leaf
- Similarly for some species the upper sunlit surface will occasionally be 1 °C warmer than the lower
- Leaves shaded from direct sunlight are usually 1 °C more than unshaded leaves
- On a clear calm night, the leaf temperature is considerably lower than air temperature
- When the sky is clear the leaf temperature will be 3 to 4.5 °C lower than the air temperature

- With increasing cloud cover the difference between air and leaf temperatures reduce
- Application: The difference between air and leaf temperature helps the farmers in adopting protective measures
    - In middle and high latitudes, frost often occurs before the air temperature drops to freezing point
    - Likewise, in summer, the heat injury to plants might be much more serious than air temperature
    - Farmers in Taiwan shade the pineapple fruit to prevent heat damage

5.5 General Effects of Air Temperature on Pests and Diseases

- Higher temperatures lead to faster development of pests and diseases
- However, as host plants also be developing faster under increased temperatures, the duration of availability of susceptible crop stages to pests and diseases may not change significantly
- Despite this, the rise in temperature will have significant effects on the incidence, severity, and spread of pests and diseases as detailed below
    - Global warming renders over-wintering of pests and diseases easier
    - Higher temperatures lead to earlier appearance and a greater number of generations of pests and diseases in a season. This results in the northward and southward spread of pests and diseases in the Northern and Southern Hemispheres respectively
    - Northward spread of (i) Cotton Cushion Scale in Europe, (ii) Cotton Camellia Scale in U.K., and (iii) Oak Moth in Central and Southern Europe are due to higher temperatures

  - Poleward extension of several crop pests in Japan in the period 1965-2000 are due to rise in temperatures
  - Migratory pests like Corn Earworm will arrive earlier at a location

- Warmer temperatures favor thermophilic fungi. Rise in temperatures increases Black Rust (Puccinia graminis tritici) and decrease Yellow Rust (Puccinia striiformis) of wheat
- Wheat and Oats become more susceptible to themophilic rusts with increase in temperatures
- Black Rust of wheat appears earlier in the wheat belt of plains of North India under increased temperatures
- In Finland and Northeastern USA, one-degree (1) centigrade rise in temperature results in earlier appearance of late blight by 4 to 7 days with an extended susceptibility period by 10 to 20 days
- Higher temperatures lead to lesser leaf-wetness duration due to higher nocturnal crop-leaf temperatures and reduce risk of incidence of many diseases
- Lower mortality of insects due to warmer temperatures leads to an increase in the population of insects
- In nature, the diversity of insect species per unit area decreases with an increase in altitude and latitude
- Thus, increase in temperatures leads to more pests attacking more crops in the temperate climatic zones
- Moderate wind blowing continuously is advantageous for the control of virus diseases in some crops. Example: Potato
- Some vectors and natural enemies of pests and diseases do better under higher temperatures. Some Aphid vectors arrive earlier or become more active. Corn Flea beetle (Chaetocnema pulicaria) is the vector for the Bacterial Blight (Erwinia stewartii) and its survival is greater at higher temperatures. The extent of survival of the vector determines the severity of Stewart's wilt infections in the following year (This can be forecast from winter temperature regimes)
- Higher temperatures cause an explosive rise in population of Ladybirds (Adalia Sp.) which are predatory on Aphids. At higher temperatures aphids become less sensitive to the

pheromones they release when under attack by enemies and thus become liable to greater predation

- Higher temperatures often offset the favorable effects of CO_2 on pests and diseases such as the cereal Aphid
- Higher temperatures change the gender ratio of some pests like thrips and reduce effectiveness of some groups/ classes of pesticides like Pyrethroids

5.6 Soil Temperature

5.6.1 Introduction to Soil Temperature

- The soil mantle of the earth is indispensable for the maintenance of plant life, affording mechanical support and supplying nutrients and water
- The soil stores energy during the warm seasons and releases it to air during the cold portions of the year
- The flux of heat into and out of the soil is a process of conduction
- However, the mechanism of transfer between the soil surface and the air is by the process of convection
- Crop plants live in two media viz., (a) Air near the ground (stems and leaves) and (b) In upper layers of the soil (roots)
- The soil temperature is critical for seed germination, root elongation, tuber development etc.
- It also effects decomposition of organic matter
- Optimum soil temperature for different crops in degrees centigrade are Wheat (15-18); Sugar beet (16); Field beans (21); Maize (20); Cotton (21); Sorghum and Soybean (25); and Tobacco (29)
- Soil temperature is also important for economic parts of plants below the soil surface. Example: The tubers of potatoes develop best at 18.0; growth reduce between 20.0-27.0 and stops at temperature of 29.0 degrees centigrade

5.6.2 General Effects of Soil Temperature on Crops

- In many instances soil temperature is of greater ecological significance to plant life Example: Oak and ash trees withstand air temperatures of -25 °C, but their finer roots succumb to cold from soil temperature of -13 °C to – 16 °C
- Soil temperature is more responsive to the local effects of insolation and topography and differ greatly from air temperature
- Many localities in polar areas and on high mountains are devoid of vegetation due to uncongenial soil temperatures
- However, in the mountain aspect and slopes where crops are grown the soil temperatures determine the growing seasons
- Soil temperatures, particularly extremes, influence the germination of the seeds, growth of roots, growth and development of the plants, and the occurrence and severity of soil borne diseases
- Extremely high soil temperatures cause destructive lesions on stems of plants and low temperatures limits plant mineral uptake
- Persistently cold soil results in dwarfed growth
- It is considered sufficient for most agricultural purposes to measure only the daily maximum and minimum temperatures, especially at great depths, because "Swift changes are encountered by the heat capacity of the soil"
- The heat capacity of the soil varies greatly according to the moisture content. Example: In a dry state, humus soils have a lower heat capacity than mineral soils. It is due to the low density of humus soils
- However, in the field organic and fine textured soils owing to their high-water holding capacity ordinarily have a higher heat capacity than coarse soils
- The rate of heat flow in the soil is determined by its temperature gradient and thermal conductivity

5.6.3 Dry Matter of Corn Vs Soil Temperature

- Dry matter accumulation in corn (maize) is significantly influenced by soil temperature, which impacts various stages of growth **(Fig:5.8)**
- Optimal temperature range for growth: Corn experiences the fastest growth rate and optimal development when soil temperature is around 27 degrees centigrade
- Average daily soil temperatures of approximately 27 degrees centigrade at a 10 cm depth are also ideal for maximizing dry matter production

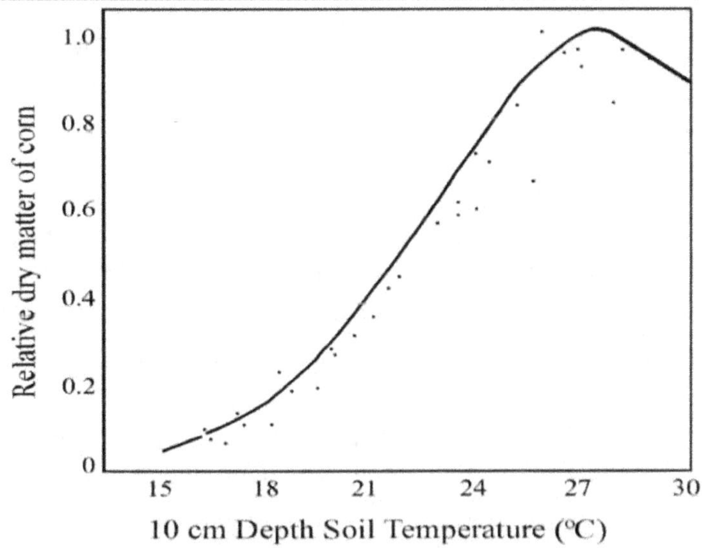

Fig. 5.8 Dry matter of corn Vs soil temperature.

Source: Unknown. Class lecture notes collection

- Impact of high soil temperatures: Elevated soil temperatures, particularly above 86°F (30°C) can reduce dry matter accumulation by hindering photosynthesis and accelerating plant development, leading to a shorter grain filling period and potentially lower yields. High soil temperatures can also intensify water stress, as plants transpire more quickly, increasing the demand for soil water and potentially accelerating drought conditions

- Effects of cold soil temperatures: Corn requires a minimum soil temperature of 50°F for germination and emergence. Temperatures below this threshold can significantly delay germination and emergence, weaken seedlings, and increase vulnerability to diseases and pests. Imbibitional chilling injury can also occur when dry seeds imbibe cold water (below 50°F), potentially damaging cell membranes and hindering germination. Wide temperature fluctuations, especially significant drops in nighttime soil temperatures, can also negatively impact emergence, particularly in sandy soils
- Soil temperature plays a critical role in corn dry matter accumulation. Maintaining favorable soil temperature conditions throughout the growing season is crucial for establishing healthy stands, optimizing growth, and maximizing yield potential

5.6.4 Soil Thermal Properties

- Soil thermal properties are very sensitive to the moisture content of soils because water has a much higher heat capacity than mineral and organic materials
- Sandy soils have steep increase of thermal conductivity with moisture
- Clay soils are less sensitive to a unit moisture change because they have much greater porosity than sandy soils
- Soil organic matter lowers thermal diffusivity because of the high porosity and consequent low thermal conductivity
- Compaction increases the thermal diffusivity by decreasing volume of the insulating pore space
- Therefore, soil heat flux is also influenced by the tilth or structure of the soil
- Plough and pack treatment produce higher thermal conductivity than ploughing, disking harrowing or rotary tillage alone

5.7 Basic Micrometeorological Terminology and Laws of Temperature for Climate Change Studies

- *Specific heat:* The amount of heat required to raise the temperature of one gram of substance by one (1) degree centigrade
- *Volumetric heat capacity*: The amount of heat required to raise the temperature of one cubic centimeter by one degree centigrade
- *Thermal conductivity*: The quantity of heat flowing per unit of time through a unit area of a plate of unit thickness when a unit temperature difference is maintained between the two opposing faces
    - The greater the thermal conductivity of the soil, the smaller the surface temperature variation and the more effective its role as a heat reservoir
    - The thermal conductivity of a soil is determined primarily by its porosity, moisture, and organic matter content
    - For a given moisture content thermal conductivity decreases from coarse to fine textured soils as porosity increases
    - The cropped fields with fine textured soils ordinarily have a higher water content which greatly increases the thermal conductivity of the soil
    - Organic matter does not transfer heat as readily as mineral soil
- *Thermal diffusivity*: A product of thermal conductivity divided by the product of density and specific heat capacity
    - It is a measure of how quickly a soil can conduct heat relative to its ability to store thermal energy. Essentially it describes how fast temperature changes propagate through the soil. A higher thermal diffusivity means heat moves through the soil more rapidly
    - The rate at which heat is transferred in a body is dependent upon the thermal conductivity of the substances plus the temperature rise that the heat produce varies with its heat capacity

- Therefore, the quotient of thermal conductivity and heat capacity is an index of the facility with which a substance undergoes temperature change and is known as thermal diffusivity
- Thermal diffusivity is crucial for understanding soil temperature variations, heat flux, and the impact of surface heating and cooling on deeper soil layers. Application: The thermal diffusivity of a soil rises with increasing moisture content. Organic matter lowers the thermal diffusivity, while packing the soil increases it

5.8 Temperature Techniques for Climate Change Mitigation and Adaptation

- Four general ways of regulating the effect of incoming energy
  - Placing an insulation layer on or near the ground surface. Example: Paper, Straw, Polystyrene, etc.
  - Heating
  - Changing the absorptivity of ground
  - Varying the air temperature by using wind machine or erection of a shelterbelt
- The thermal properties of the ground can be modified
  - By increasing the absorptivity of the ground
  - By changing the thermal conductivity by cultivation, rolling, and irrigation
  - By altering the heat capacity through adding or draining water
  - By varying the rate of evaporation
  - Examples: Removal of weeds, regulating soil moisture, placing mulch, screens, sand on the ground surface

5.9 Addressing Climate Change Challenges of Temperature

- The individuals, families, communities, governments, organisations, etc., have the best sciences available to plan for and implement climate-smart and resilient practices for temperature increases due to climate change. Two critically important approaches which are available to managing climate resilience are mitigation and adaptation. These strategies are inter-related and must be pursued simultaneously as both can potentially improve resilience to the changing climate.

Chapter 6

Climate Change and Rainfall, Dew, and Fog

"From the heavens, nourished by the Sun and driven by cosmic rhythm (rita), rain descends as a divine blessing, sustaining all life on Earth. This sacred act overseen by deities like Parjanya and Varuna represents the renewal and flow of cosmic and moral order. Thus, rainfall mirrors life's cyclical nature: evaporation, condensation, and precipitation become a metaphor for creation, sustenance, and dissolution"

6.1 Introduction to Rainfall

- The hydrological cycle, also known as the water cycle, is the continuous movement of water on, above, and below the surface of the Earth **(Fig:6.1)**
- Rainfall is one of the basic inputs of hydrological cycle and thus requires understanding of its dynamics, trends and variability
- The key processes involved in hydrological cycle are
- Evaporation and transpiration: Driven primarily by the solar radiation (Energy), water changes from liquid state to gas state and rises into the atmosphere, known as evaporation. This occurs from oceans, seas, rivers, lakes, and other bodies of water while transpiration is the release of water vapor from plants
- Condensation: As water vapor rises and cools it transforms back into liquid water droplets or ice crystals forming clouds

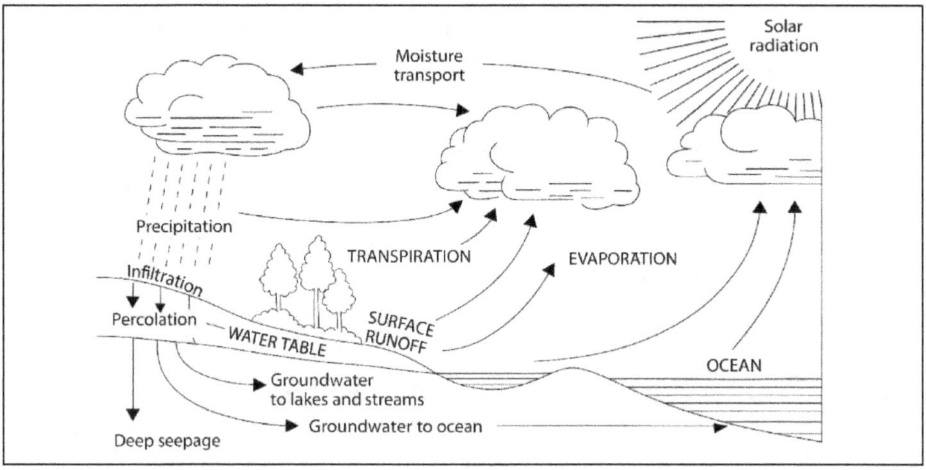

Fig. 6.1 The Hydrological cycle.
Source: Unknown. Class lecture notes collection.

- Precipitation: When the water droplets or ice crystals in clouds become heavy enough, they fall back to the Earth as precipitation/ condensation (rain, snow, hail, dew, fog, etc.)
- Runoff: When precipitation falls on land, it flows over the surface as runoff, making its way to rivers, lakes, and eventually to the seas and oceans
- Infiltration and groundwater: Some precipitation soaks into the ground, becoming soil moisture or groundwater, which flows slowly through the subsurface or be drawn up by plants or used by humans through wells
- Essentially, the hydrological cycle describes how water circulates among the oceans, atmosphere, and land constantly changing states (solid, liquid, and gas) and locations
- This cycle is fundamental to the Earth's climate, agroecosystems, and the availability of freshwater resources
- The aspects of the hydrological cycle, involving the soil are important in climate change studies and agricultural meteorology
- The reservoir of water in the soil permits transportation to continue and plants to survive between rains

- The precipitation is divided into components according to how the water returns to the atmosphere
- A certain fraction of the precipitation is intercepted by the plant canopy and evaporated without reaching the soil surface
- Rainfall plays a pivotal role in defining the climatology of a region
- Shifting in rainfall pattern results in shifting/ changing the dates of sowing of crops

6.2 Trends, Sources, and Projections

- Climate change is affecting the intensity, duration, amount, patterns, etc., of rainfall
- Heavy precipitation events have become more frequent over most areas during the twentieth century
- The areas affected by drought have increased in many regions since 1970
- Intense tropical cyclone activity has increased in some regions since 1970
- Observations show that changes are occurring in the amount, intensity, frequency, and type of precipitation/ rainfall
- These aspects of precipitation generally exhibit large natural variability, in addition to El Nino, La Nina, and changes in atmospheric circulation patterns such as the North Atlantic Oscillation that have a substantial influence
- Both the range and intensity of ENSO events are much higher since 1950
- Due to global warming ENSOs have the most significant effect on climate change

- Pronounced long-term trends from 1900 to 2005 have been observed in precipitation amount in some places viz., significantly wetter in eastern North and South America,
- northern Europe and northern and central Asia, but drier in the Sahel, southern Africa, the Mediterranean and southern Asia
- More precipitation is falling as rain rather than snow in northern regions
- Widespread increases in heavy precipitation events have been observed even in places where total amounts have decreased
- These changes are associated with increased water vapor in the atmosphere arising from the warming of the world's oceans, especially at lower latitudes
- Each additional one (1) degree centigrade temperature intensifies extreme rainfall by 7 per cent
- More global warming results in heavier rainfall thereby changes in the hydrological cycle

6.3 General Effects of Rainfall on Crops

6.3.1 Effects of Water on Plant Physiological Processes

- Major constituent of physiologically active tissue
- A reagent in photosynthesis
- A solvent in which salts, sugars, and other solutes move from cell to cell and organs
- Essential for plant turgidity which is necessary for cell enlargement and growth
- Needed for transpiration
- Water deficiencies not only reduce crop yields but also change the patterns of growth
- In general, effective rooting depth decreases as the soil moisture level increases
- Roots developed under limited moisture conditions are finer and have more and longer branches than those developed under unlimited soil moisture conditions

- Therefore, frequent irrigation often leads to shallow and horizontally spread root systems which are a disadvantage when dry conditions occur
- The ratio of root to shoot is usually increased by water stress
- Also, under water stress leaf area is often reduced but leaf thickness is increased
- The quality of growth, flowering, seed formation, and economic yield are all influenced by moisture conditions

6.3.2 Variation in Crop Water use during Growing Season

- Crop water use, also known as evapotranspiration (ET) is the total amount of water transferred from the land surface to the atmosphere by evaporation from the soil surface and transpiration from plants
- Growing Season Dependence
    - o During germination, emergence, and vegetative growth average crop water use occurs
    - o Crop water requirements change throughout the growing season, peaking during reproduction (seed set) **(Fig:6.2)**

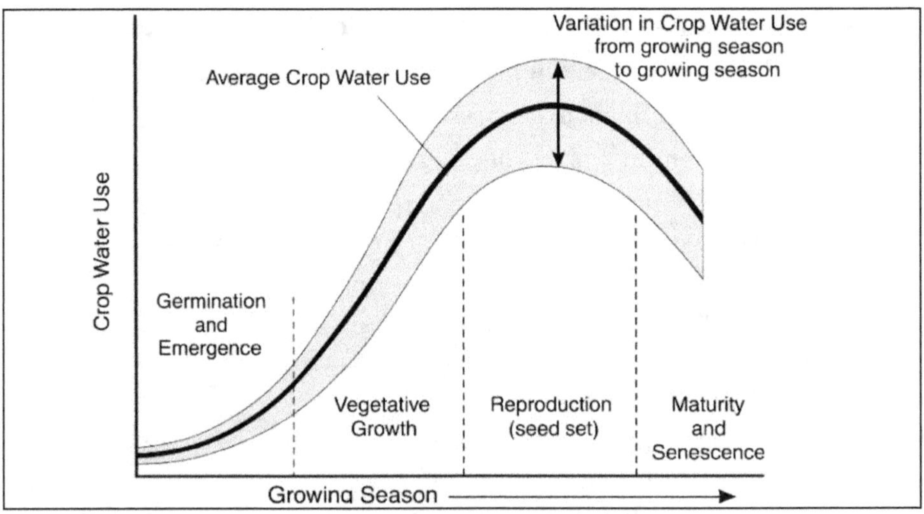

Fig. 6.2 Variation in crop water use during growing season.
Source: Unknown. Class lecture notes collection

  - o During maturity and senescence stages the crop water use decreases
  - o In addition, variation in water use occurs from growing season to growing season

- Influencing Factors: ET is influenced by numerous factors, including
  - o Climatic conditions: Temperature, sunshine (Solar radiation), wind speed and humidity
  - o Crop characteristics: Crop type, growth stage, leaf area, and root depth
  - o Soil factors: Soil moisture content, type, and water-holding capacity
  - o Management practices: Planting dates, densities, irrigation method, and crop residue
- Significance: Understanding crop water use is crucial for
  - o Efficient irrigation management: Scheduling irrigation to match the crop's needs at different growth stages, minimizing water waste, and maximizing water use efficiency
  - o Optimizing crop production: Ensuring adequate water availability, especially during critical growth periods to maximize yields
  - o Developing adaptation strategies: Breeding drought-tolerant crops and adopting water-saving practices to cope with climate change and water scarcity
- Crop water use is a dynamic process governed by a complex interplay of environmental, crop, and management factors
- Optimizing water management during the growing season is critical for both yield and water conservation

6.3.3 Water Use Efficiency and Climate Change

- Water Use efficiency (WUE) in crops refers to the amount of biomass or grain yield produced per unit of water consumed i.e., water used **(Fig: 6.3)**

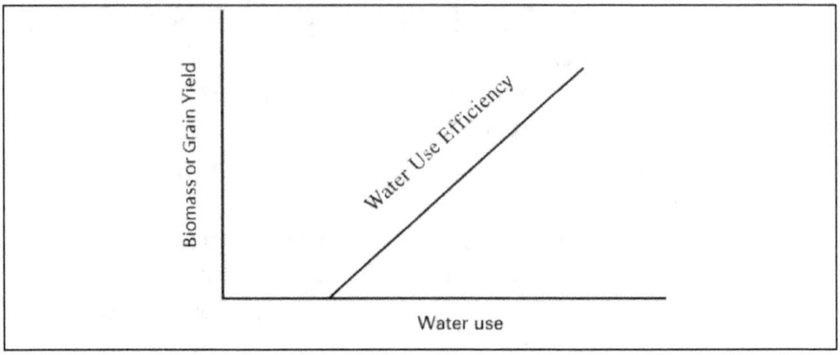

Fig. 6.3 Water use efficiency as a function of water use and biomass or grain yield.
Source: Unknown. Class lecture notes collection

- It is a measure of how effectively a plant converts water into valuable plant material
- A high WUE indicates a plant is efficient at producing yield with less water
- WUE can be calculated by dividing biomass or grain yield by water use, often estimated as evapotranspiration (ET)
- Definition is essentially a ratio: WUE = Biomass or Grain Yield / Water Use
- Importance: In the context of agriculture, WUE is crucial for optimizing water resources especially in water-limited environments
- Factors Influencing WUE
- Several factors influence WUE, including photosynthesis, transpiration, stomatal conductance, and the plant's ability to access and utilize water from the soil
    - Leaf-level WUE: At the leaf level, WUE can be influenced by factors like CO_2 concentration, temperature, and water availability
    - Whole-plant WUE: At the whole-plant or canopy level, WUE is affected by factors like soil evaporation, transpiration, and the overall growth pattern of the crop
    - Improving WUE: Strategies to improve WUE include optimizing irrigation practices, selecting drought-tolerant

- varieties, and managing canopy architecture to enhance light interception and water uptake
- Climate Change: WUE is an important consideration in the context of climate change, as it can help identify crops that are more resilient to water scarcity

6.3.4 Relationships between Wheat yields and Seasonal Rainfall in Israel (Arid) and France (Humid)

Israel (arid)

- Wheat yields are highly dependent on rainfall, particularly during critical growth stages
- Since Israel is an arid region wheat yields steadily increase as seasonal rainfall increases to around 500 mm **(Fig: 6.4)**

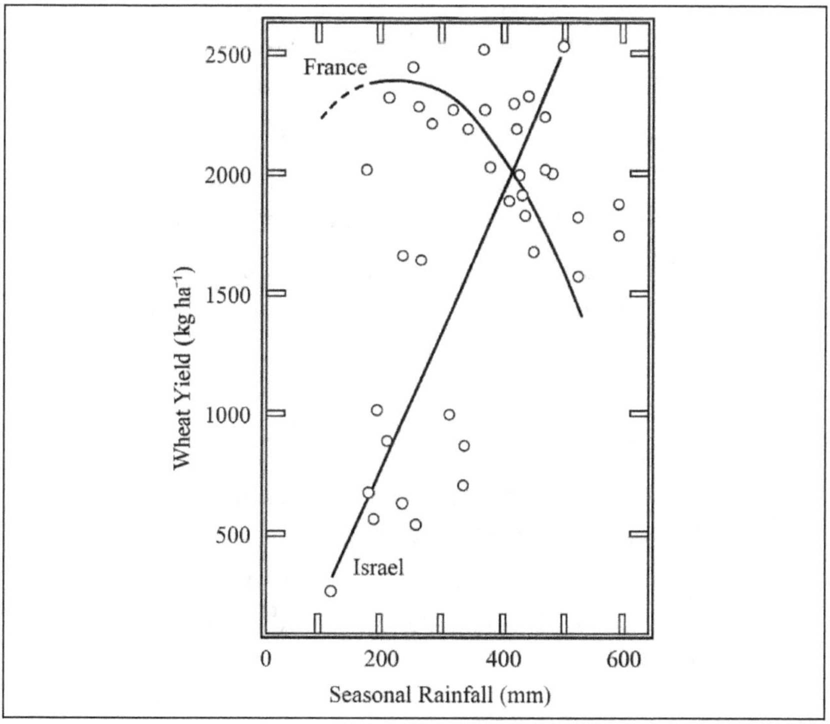

Fig. 6.4 Relationship between wheat yields and total seasonal rainfall in Israel (Arid) and France (Humid).

Source: Unknown. Class lecture notes collection

- Rainfall variability, characterized by droughts and decreased winter rainfall, has a significant impact on wheat production
- Strategic water management practices, including desalination and efficient irrigation are crucial for sustaining wheat production in Israel's arid climate

France (humid)

- Wheat cultivation thrives in France's temperate climate under adequate rainfall
- However, excessive rainfall, especially during crucial phases, can negatively impact yields through problems like lodging and fungal infections
- Since France is a humid region the wheat yields start decreasing from a seasonal rainfall of 300 mm
- Dry periods and insufficient rainfall, particularly in central and northern regions, also threaten yield if they persist
- Timely and adequate rainfall throughout the growing season is critical for optimal wheat development and yield in France
- In arid Israel, water availability is the primary driver of wheat yield, with rainfall patterns dictating success or failure
- In humid France, the balance of rainfall is key. Adequate moisture is crucial, but extremes (both excessive and insufficient) can be detrimental to yields

6.3.5 Relationships among Wind Erosion, Rainfall, and Vegetation

- The relationships among wind erosion, rainfall, and vegetation are complex and interconnected **(Fig:6.5)**

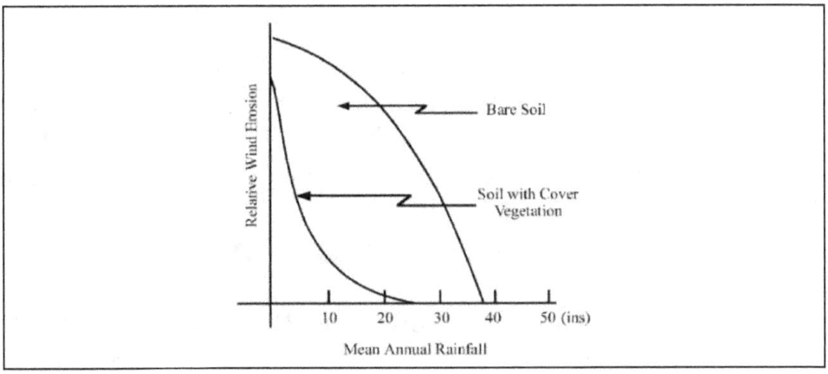

Fig. 6.5 Relationships among wind erosion, rainfall, and vegetation.
Source: Unknown. Class lecture notes collection

- Wind erosion and vegetation
    - Wind erosion is less when soil is covered with vegetation as compared to bare soil
    - Vegetation, especially deep-rooted plants, helps anchor soil and prevent it from being carried away by wind
    - Plant canopies reduce wind speed near the ground, further lessening the erosive force of thc wind
    - Denser vegetation provides more effective wind erosion control
- Wind erosion and rainfall
    - Rainfall can inhibit wind erosion by increasing soil moisture, binding soil particles, and making the soil less susceptible to being blown away by wind
    - However, heavy or intense rainfall can lead to water erosion and soil compaction, potentially increasing the vulnerability to subsequent wind erosion once the soil dries out
    - Water erosion from a bare soil is more as mean annual rainfall increases
- Rainfall and vegetation
    - Rainfall is essential for vegetation growth and development
    - Changes in rainfall patterns, such as prolonged droughts or increased aridity due to climate change, can negatively impact vegetation growth, leaving the soil surface more exposed and susceptible to both wind and water erosion

- Healthy vegetation acts as a crucial buffer against both wind and water erosion, while appropriate amounts of rainfall, supports vegetation growth and helps to bind the soil and prevent wind erosion. However, imbalances in rainfall patterns can compromise this natural protection and exacerbate erosion risks

6.4 Evaporative Power of Air (EPA)

- Evaporative Power of Air (EPA) is the capacity of air to desiccate moist surfaces given 100 per cent opportunity
- Global warming increases evaporative demand
- As a result, irrigation demand increases as temperature rises because of increased evapotranspiration
- On the other hand, the Evaporative Power of Air (EPA) decreases due to reduced bright hours of sunshine brought about by higher aerosol concentration due to climate change
- Increased cloudiness and/or higher aerosol concentration reduces evaporation
- Which mechanism between "Global warming" and "Reduced bright sunshine hours" operates first decides the EPA
- C_4 crops (maize, sorghum, pearl millet, etc.) require less water per gram of carbon assimilated than C_3 crops (rice, wheat)
- Therefore, the above C_4 crops are preferred in dry land areas

6.5 Model Simulated and Model Projections of Indian Summer Monsoon Rainfall (ISMR) in the Global Warming Scenario

- The South Asian region with high human population and monsoon related economies are highly vulnerable to changes in monsoon precipitation due to climate change. Therefore, the possible impacts of climate change on Indian Summer Monsoon (ISM) and its variability assume significance
- Changes observed in Indian Summer Monsoon during 20th century
- Extremes in precipitation

    - Significant rising trends in the frequency and magnitude of extreme rain events during 1951 – 2000
    - Significant decreasing trends in the frequency of moderate events over central India during monsoon seasons during 1951 – 2000
    - Seasonal mean rainfall does not show significant trends. It is because of the contribution from the increasing heavy events that are offset by decreasing moderate events
    - It is projected and a possibility that in future there may be substantial increase in hazards related to heavy rainfall over central India
    - While an increase in rainfall is shown over regions with fast industrial growth, a decreasing trend in rainfall over high mountain ranges which are influenced by intense deforestation
    - Inter decadal variability in the frequency of extreme rain events has also been reported
- Extremes in temperature
    - Indian annual "Mean, max, and min temperature" have shown significant warming trends (1901-2007)
    - After the mid-1990s, nighttime temperatures started increasing at twice the rate of daytime temperatures
    - Widespread increasing trends are observed in the frequency of hot days and hot nights, while that of cold days and cold nights a widespread decreasing trend
- Cyclonic disturbances
    - Decreasing trends are observed in cyclonic disturbances over Bay of Bengal and North Indian Ocean
    - Significant increase in the frequency and duration of low-pressure areas for the recent decades has been reported
- Active and break spells
    - No significant trends in either the days of active or break events were noted
- Summary
    - An increase in heavy rainfall events and a decrease in moderate and low rain events in particular during the 2nd half of the 20th century

  - The extreme rain events suggest an increase in the risk of major floods in general and over central parts of India in particular
  - Some changes in intra-seasonal behavior of monsoon rainfall have been reported
  - Increase in overall Indian Summer Monsoon Rainfall over India is projected
  - In particular, over the Arabian Sea, West Coast, and North West India, i.e., regions with heavy rainfall will continue to get heavier rainfall
  - Increase in rainfall could be due to projected intensification of the heat low over Northwest India
  - A major cause for the increase in precipitation may be due to increase in atmospheric water vapor over the land areas advected from the warmer Bay of Bengal and Arabian Sea
  - Temperatures have been increasing over the Indian sub-continent during the recent decades
  - Night time temperatures have started increasing at twice the rate of daytime temperatures
  - Widespread increasing trends in temperature are reported, with the possibility of increase in severe heat waves
- IPCC Projections of Indian summer monsoon
  - The IPCC provided outputs of 25 coupled ocean atmosphere models
  - The studies project an increase in precipitation of about 8 per cent over South Asia and 10 per cent over East Asia
  - The increase over South Asia could be due to the increase in atmospheric water vapor
  - While increase over East Asia could be due to the intensification of North Pacific Sub-Tropical High and Meiyu Changma Baiu front (MCBF)

6.6 Maximization of Rainfall usage

- Unlike irrigation, rainfall is non- monetary
- It can be utilized maximum by maximizing the direct use of rainfall in crop production

- Harvesting surface runoff from rains through construction of agroclimatologically designed, cost effective on farm reservoirs, and reusing the water so collected in drought situations
- Augmenting recharge of ground water and strategically using the same to combat drought

6.7 Introduction to Dew

- The distribution of precipitation/ forms of condensation are seldom uniform in space and are never uniform in time. Dew and fog are highly significant in agriculture
- After a sunny day, with little or no wind, and an evening with a clear sky and high humid the next morning might show the grass fields, trees, and other vegetation or objects exposed to the sky covered with small droplets of various sizes
- This phenomenon of the formation of water droplets is "Dew formation"
- Dew is the deposit of water droplets on objects the surface of which is sufficiently cooled, generally by nocturnal radiation, to bring about the direct condensation of the water vapor from the surrounding air
- Whenever the air temperature around different non hygroscopic objects such as leaves, flowers, grass, windows or other exposed object falls below the dew point temperature water droplets of various size will form on their surface

6.7.1 Factors affecting Dew Formation

- Dew is deposited because the air near the surface is cooled. To understand the factors that determine dew formation, the understanding of factors that favor or inhibit the cooling of such surfaces is necessary
- *Clear skies:* During the night all surfaces lose heat by radiation to the atmosphere, especially if the sky is clear. This may cause the temperature of the radiating bodies to be as much as 2,3,4, or 5 °C less than that of the surrounding air

- *Degree of exposure of radiating surfaces:* An isolated object losing heat by radiation during the night, may do so more easily and more quickly than a collection of objects which may mutually intercept some radiation emitted by nearby objects. This is observed readily by looking at the amount of dew deposited on grass surfaces in open areas and in protected areas
- *Low winds:* When there is no or little wind, <3m $\check{s}^{-1}$, the dew deposit may increase. At low wind speeds, the air in contact with objects is being renewed by air from similar levels of height. This causes the formation of dew to be continuous and abundant. When wind speeds become greater than 5 or 6 m $\check{s}^{-1}$, no dew is formed because the air changes rapidly. Then some evaporation may take place and the temperatures of the surfaces in contact with the air may not become as low as the dew point temperature
- The role of higher/greater density of cool air
    - It can be observed on some mornings that, in orchards, the tops of the trees are dry or closure to dry while the soil between the trees may be covered in dew
    - The upper parts of the trees are more exposed to radiation losses and cool more rapidly
    - As a result, the air descends between trees as dew forms because cooler air has a greater density
    - The air near the top of the trees is replaced by slightly warmer air and the process continues
    - Thus, the cooling of upper leaves and the deposition of dew on them is retarded or prevented
    - In some villages of mountain regions dew is formed in the lowest lying areas
    - It is observed that dew is formed in the valley bottom while there is little or nothing on the slopes
    - This phenomenon is similar to that of orchards as detailed above
    - When the air over the slopes cools, it becomes heavier, descends to the valley floors and continues to cool by radiation eventually giving rise to dew formation

- o The factors that cause the deposit of dew at any place are complex and may interact
- o These include solar radiation, temperature, humidity, micro relief, soil colour, vegetative cover, etc.
- o Plants may use water in its vapor form when RH is both high and low
- o Thus, some crops (Bengal gram) may obtain water through aerial parts as well
- o Plants explore all different sources of water under adverse conditions

6.8 General Effects of Dew on Crops

- In tropical and sub-tropical regions with low rainfall, but close to seas, rivers or lakes, the high day time temperatures cause high evaporation and water vapour remain in the ambient air
- During night when temperatures decrease, condensation occur covering crops with dew
- The annual totals of dew are rarely more than 50mm
- The lower species of plants, such as algae and fungi use these resources of water very efficiently
- Certain higher species of epiphytic plants such as orchids are well adapted to have a morphology that allows them to make good use of dew
- Other higher species of plants are also well adapted. Example: Prosophis
- Crops like tomato, potato, tobacco etc., use dew when they are grown in areas of between 20 and 50 mm of dew
- Some fodder crops grow well using dew
- Under changing climate scenarios dew play an important role in production of crops

6.8.1 Disadvantages of Dew

- Development of bacteria and fungi on crops
- Animal diseases when dew wet fodder is fed

6.8.2 Application

- Knowledge of these characteristics helps to define processes to use dew more efficiently for plant growth or devise frost protection measures
- Grapevines should be planted in hollows, 1.5 to 2m deep and 3 to 4m in diameter so that cool air draws into the hollow and dew deposits are favoured. Grapevines are grown on natural or artificial hollows so that when air cools, it drains into these hollows and deposits dew when it cools below the dew point. Stone walls prevent the sides from caving in
- For wind protection walls constructed around apple orchards also help in drawing cool air and deposition of dew

6.9 Fog

6.9.1 Introduction to Fog

- Fog is low clouds or clouds at soil level
- Suspension of very small, usually microscopic water droplets in the air generally reducing horizontal visibility at earth's surface to less than 1 km
- Haze is when horizontal visibility is between 1 km and 5 km
- Fog is same as dew from agrometeorological point of view
- Dew forms on surfaces. Fog occurs on lower atmosphere
- When air cools, it becomes denser and heavier and flows to lower lying places
- Fog forms in these places with a thickness of a few meters to a few hundreds of meters. It depends on (a) Intensity of cooling and (b) Water vapour content of air
- In case of fog the lower layers of the atmosphere need to be cooled to the value of dew point

6.9.2 Conditions for Fog Formation

- Heating during the day, clear skies, light wind < 3m s^{-1},
- Thermal inversion and high humidity
- Usually forms during the night and early morning

- Forms more on land than on water
- When hot and humid air mass flows over a relatively cold surface advection fog forms
- Adiabatic cooling of air when it moves upward, over a slope, leads to condensation and orographic fog is formed
- Katabatic fog is formed when cold air descends along the mountain slopes towards the valley bottom
- Fog is formed by water evaporating from rivers or lakes or by water transpired by vegetation into cool air, which facilitates its condensation
- This is very characteristic over rivers, valleys, lakes, behind dams, etc.

6.10 Uses of Fog in Agriculture

- When fog moves up against the slope of a hill, it will leave water on all non hygroscopic objects on its way, more than 10 cms tall
- The presence of vegetation enhances the capture of water
- Assessment of Pests and diseases
- Examples
  - Mildew on grapevines T = 22 °C R.H. = 70%
  - Cercosposa on sugar beet T = 28 °C R.H. 90%
  - Rust in wheat T = 20 °C R.H.95%
  - Meteorological conditions help in probability of occurrence of pests and diseases and to take up protection measures

6.11 Occult Precipitation

- Definitions
  - Precipitation occur at a location by precipitation/ condensation processes that would normally go unrecorded by a standard rain gauge
  - Occult precipitation is the supply of water to soil or vegetation that is not by way of straight forward rain and so is not measured by conventional rain gauges

- Fog, dew and cloud water are inputs for occult precipitation
- Occult precipitation is a natural resource that is not much used but is of potential importance in some of the least endowed regions of the world
- Under climate change scenarios occult precipitation play significant role in crop production

6.12 Basic Micrometeorological Terminology and Laws of Precipitation/ Condensation for Climate Change Studies

- Fick's Law of Diffusion
  - Definition: Movement of substances from an area of high concentration to an area of less concentration because of the random motion of individual molecules
  - Fick's Law describes this process which is called as "Diffusion"
  - In simpler terms, imagine a drop of ink in water. The ink spreads out until it is evenly distributed throughout the water, and this spreading is diffusion governed by Fick's Law
- Mass transport: This is referred to as "Diffusion". The spreading of mass from highly concentrated areas to less concentrated areas is "Diffusion" or "Mass transport" which is described using Fick's First Law
- Mass transfer/ Mass transport
  - This is the net movement of mass from one location usually meaning stream, phase, fraction, or component to another. Mass transfer occurs in many processes such as absorption, evaporation, drying, precipitation, membrane filtration, distillation, etc.
  - Mass transfer is used by different scientific disciplines for different processes and mechanisms. The phrase "Mass transfer" is commonly used in engineering for physical processes that involve diffusive and convective transport of chemical species (Substances) within physical systems

- o Renold's number and Schnidt number characterize diffusional properties of flowing media. Example: The evaporation of water from a pond to the atmosphere

6.13 Rainfall Techniques for Climate Change Mitigation and Adaptation

6.13.1 Management of Cyclones, Floods, Droughts, and Different Forms of Condensation

- Preparedness for impending cyclones, floods, and droughts, and different forms of condensation that negatively impact agriculture refer to the plan of action needed to minimize loss to human lives, damage to property and agriculture. The effectiveness of disaster preparedness ultimately depends on the effectiveness of planning and response at a district or local government level
- Preparedness in agricultural systems include early harvesting of crops at physiological maturity, safe storage of harvest, cultivation of short duration crops or varieties which are not easy grain shedders, cultivation of spreading type of crops or varieties, alternative cropping pattern in areas affected, reclamation and treatment of soils, maintenance of ecological balance through large scale plantation, growing vegetables, nurseries and kitchen gardens, etc.

6.14 Addressing Climate Change Challenges on Rainfall and other Forms of Condensation

- The individuals, families, communities, governments, organizations, etc., have the best sciences available to plan for and implement climate-smart and resilient practices. Two critically important approaches which are available to managing climate resilience are mitigation and adaptation. These strategies are inter-related and must be pursued simultaneously as both can potentially improve resilience to the changing climate.

Chapter 7

Climate Change and Sea Level Rising and Extreme Events

"Oceans represent the primordial waters (apah), the boundless source of creation and sustenance, both external and internal within the heart. They symbolize the vastness of consciousness and wisdom, the destination where all rivers (individual souls) ultimately merge with the divine (Brahman). From these depths emerge the elixir of immortality (amrit) and divine blessings, yet they also embody impermanence, reflecting the continuous cycle of life, death, and renewal"

7.1 Introduction to Climate Change and Sea Level Rising

- An important manifestation of changes in the climate system due to climate change is the warming of the world's oceans
- Around 67 per cent of the Earth is oceans
- During the onset of the Industrial Revolution and the twentieth century the anthropogenic effects from greenhouse gases have led to accelerated icesheet melting and an increased rate of sea-level rise
- Warming of oceans and ocean acidification proves critical and challenging for the present and future generations

- Sea level rise has profound and multifaceted effects on coastal regions and beyond
- These impacts include increased flooding, coastal erosion, saltwater intrusion into freshwater sources, and displacement of populations
- Rising sea levels lead to more frequent and severe flooding during high tides and storms inundating low-lying areas and damaging property
- Furthermore, sea level rise exacerbates the impacts of storms threatening infrastructure and ecosystems
- Salinization of soil and water can reduce agricultural productivity impacting food security and livelihoods
- Seawater contaminates freshwater sources like aquifers and agricultural land impacting drinking water and irrigation
- Rising sea levels can alter ocean currents and temperatures affecting marine ecosystems and the species that depend on them
- Damage to property, infrastructure, and tourism-related industries can result in significant economic losses for coastal communities

7.2 Causes of Sea Level Rising

- The three main causes of sea level rise are: Thermal expansion, melting glaciers and ice sheets, and changes in land water storage. Essentially, as the Earth warms due to climate change the ocean expands, ice melts adding more water, and changes in water stored on land contribute to the rise
- Thermal expansion: As the ocean absorbs heat from the atmosphere (caused by greenhouse gases) the water expands in volume. This is known as 'Thermal expansion" and it is a significant contributor to rising sea levels
- Melting glaciers and ice sheets: Glaciers and ice sheets located on land store vast amounts of frozen water. As global temperatures rise this ice formations melt at an accelerated rate, and the meltwater flows into the oceans increasing the overall volume of water

- Changes in land water storage**:** This includes changes in water stored in reservoirs, aquifers, lakes, and soil moisture. Water that was previously stored on land may be transferred to the oceans either through natural processes or human activities like groundwater pumping

7.3 Trends, Sources, and Projections

- Other factors contributing to sea level rise over the last decade include a decline in mountain glaciers and ice caps (0.77 ± 0.22 mm per year)
- Approximately two-thirds of the global coastline have a projected regional relative sea level rise within ±20 per cent of the global mean increase
- According to IPCC 2021 reports, the global mean sea level increased by 0.20 m between 1901 and 2018
- In addition, the global ocean temperature rose by 0.10°C from the surface to 700 m depth from 1961 to 2003
- More than 90 per cent of the energy accumulating in the climate system between 1971 and 2020 has accumulated in the ocean
- Therefore, the global sea level is projected to rise during the 21st century at a greater rate than during 1961 to 2003
- Satellite altimetry data shows that since 1993 sea levels have been rising at a rate of around 3 mm per year significantly higher than the average during the previous half century
- The present rate of sea level rise is 3 mm per year
- According to IPCC 2021, in 2011–2020 annual average Arctic Sea ice area reached its lowest level since at least 1850
- At present, late summer Arctic Sea ice area is smaller than at any time in at least the past 1000 years
- It is very likely that regional mean relative sea level rise will continue throughout the 21st century except in a few regions with substantial geologic land uplift rates

- The global nature of glaciers retreats with almost all the world's glaciers retreating synchronously since the 1950s is unprecedented in at least the last 2000 years
- Coastal tide gauge measurements confirm this observation
- Satellite observations available since the early 1990s provide more accurate sea level data with nearly global coverage
- Land temperatures are rising at historic heights, simultaneously, the ocean temperatures also continue to climb
- The IPCC projects that global sea level will continue to rise throughout the 21st century and beyond driven by factors such as the thermal expansion of warming seawater, the melting of glaciers, and ice sheets and changes in land water

7.3.1 Regional Variability

- Sea level rise is not uniform globally. Regional and local factors such as ocean currents and land elevation changes can cause local sea level rise to differ from the global average
- Rising sea levels increase the frequency and severity of extreme sea level events like storm surges and coastal flooding, particularly in low-lying coastal areas
- Many coastal areas could experience events that previously occurred once a century becoming more frequent occurrences by 2050

7.3.2 Long-Term Impacts

- Even if global warming is limited, sea level rise will continue for centuries due to the slow response of oceans and ice sheets to warming
- According to some estimates sea levels in 2300 might exceed 3.5 meters (11.5 feet) under very high emission scenarios, while achieving the Paris Agreement goal would likely limit the rise below 1 meter (3.3 feet)
- Coastal communities face various long-term challenges including erosion, salinization of agricultural land, and displacement of populations

7.3.3 Latest IPCC Reports on Sea Level Rising

- According to IPCC the global mean sea level is rising relative to 1900. It is due to human-induced global warming **(Fig:7.1)**

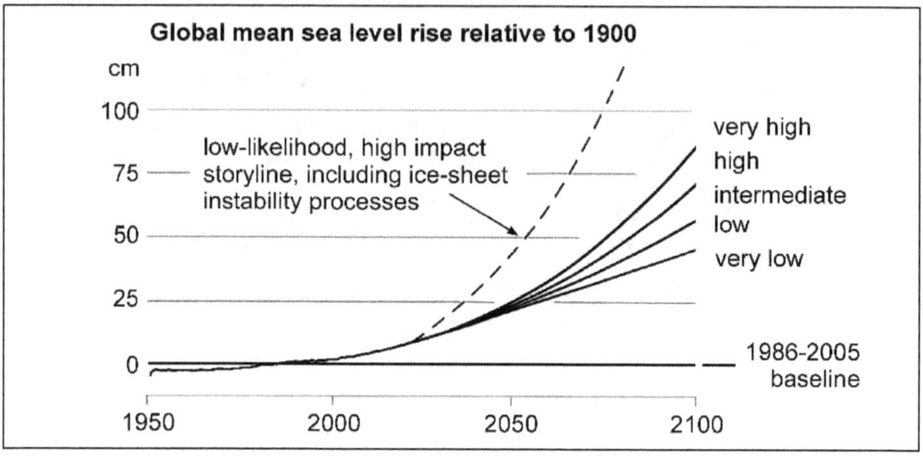

Fig. 7.1 Global mean sea level rise relative to 1900.
Source: IPCC, 2021, 2022, and 2023

- Past and current trends: GMSL rose significantly during the 20th century and is currently rising at more than twice the 20th-century rate. The IPCC has high confidence that this rise is unprecedented in at least the last 3,000 years
- Causes: The primary causes of GMSL rise are the thermal expansion of ocean water (As ocean warms, it expands) and the melting of land-based ice (glaciers and ice sheets). While thermal expansion was dominant in the 20th century, ice loss from glaciers and ice sheets is now a larger contributor
- Future projections: Sea levels are projected to continue rising for centuries under all emissions scenarios. The extent of the rise depends heavily on future greenhouse gas emissions
    - Even with significant emission reductions, GMSL is projected to rise by around 30-60 cm by 2100
    - With continued high emissions, the rise could be significantly higher, reaching around 60-110 cm by 2100, and potentially exceeding 2 meters by 2100 and 5 meters by 2150 if rapid ice sheet loss occurs

- Impacts: Rising sea levels increase the frequency and intensity of extreme sea level events (like storm surges and high tides) leading to increased coastal flooding, erosion, and potential inundation of low-lying coastal areas and islands. This poses risks to crops, coastal ecosystems, infrastructure, freshwater resources, and human health
- Uncertainties: There is still uncertainty surrounding the future behavior of the Greenland and Antarctic ice sheets, especially under higher emission scenarios. This uncertainty makes it challenging to project the full range of potential sea level rise, particularly the risk of rapid ice sheet collapse and its contribution to GMSL rise

7.4 Carbon Assimilation by Oceans

- Through various physical, chemical, and biological processes the ocean absorbs a substantial amount of carbon dioxide from the atmosphere
- Carbon dioxide dissolves into seawater at the surface, driven by the difference in partial pressures between the atmosphere and the ocean
- The ocean plays a vital role in carbon assimilation, acting as a significant carbon sink and helping to regulate atmospheric carbon dioxide levels
- The absorption of excess carbon dioxide by the oceans lead to a decrease in pH, a process known as "Ocean acidification"

7.5 Mitigation and Adaptation Strategies

- Strong mitigation efforts (reducing greenhouse gas emissions) are crucial to limit the extent of future sea level rise. Additionally, adaptation strategies are essential to manage the unavoidable impacts, including coastal protection (seawalls, dikes), ecosystem restoration, climate-smart planning, and even relocation in some cases

- Deep and sustained reductions in greenhouse gas emissions are critical for limiting the amount and rate of sea level rise
- Adaptation strategies are crucial for coastal communities to prepare and respond to rising sea levels including implementing protective measures like seawalls, accommodating coastal changes or considering managed retreat

7.6 Extreme Weather Events

7.6.1 Introduction to Extreme Weather Events

- Basics of Extreme Events
    - Throughout human history natural disasters have played a major role in the economic development and survival of humanity
    - With the increasing threat of global warming there is a drastic increase in the number of catastrophic natural disasters and extreme events
    - Every year irregular and extreme weather events are reported worldwide
    - Climate change is affecting every inhabited region across the globe with human influence contributing to many observed changes in weather and climate extremes
- Definition: Extreme weather events are unusual, severe, and unseasonal weather conditions that deviate significantly from natural climate resulting in significant damage to ecosystems, infrastructure, and human lives

7.7 Trends, Sources, and Projections

- 2022: The "State of the Global Climate" report showed record levels of heat-trapping greenhouse gases and melting glaciers
- 2024: A total of 605 extreme weather events occurred with 148 classified as unprecedented
- 2024 also saw unprecedented global temperatures exceeding in regions the 1.5°C threshold set by the Paris Agreement

- Nine of the top ten years of extreme one-day precipitation events have occurred since 1995
- Hurricanes are becoming more powerful due to warmer ocean temperatures
- There are regional variations with some areas experiencing more frequent and intense floods while others face more frequent and intense droughts
- Heat waves, heavy rainfall, droughts, and wildfires are becoming more common and severe, while cold spells are decreasing in some regions
- These changes are largely attributed to human-caused greenhouse gas emissions and the resulting warming of the planet
- These events are not only becoming more frequent but also more intense leading to significant impacts on human health, agroecosystems, and infrastructure

7.8 Increased Heat Waves

- Warmer temperatures lead to more frequent and intense heat waves both on land and in the oceans
- Even a small increase in average temperatures can significantly amplify heat waves
- For example, the Sahel region in Africa experienced a record-breaking heatwave in April 2024 with temperatures exceeding 48°C (118°F)
- Studies project that heat waves will become more frequent and severe in all regions of the world throughout the 21st century
- Heatwaves are significant contributor to economic losses and fatalities and are becoming more frequent and intense

7.9 More Intense Rainfall and Floods

- Generally heavy daily precipitation events that lead to flooding have increased

- Increasing trends are reported as well for heavy precipitation events
- The frequency and intensity of heavy precipitation events have increased since the 1950s over most land areas for which observational data is sufficient for trend analysis and human-induced climate change is the main driver
- A prominent indication of a change in extremes is the observed evidence of increases in heavy precipitation events over the mid-latitudes in the last 50 years even in places where mean precipitation amounts are not increasing
- Spatio - temporal mapping of flood inundation is important for policy makers towards flood preparedness, raising awareness, designing the mitigation strategies, and implementation of flood management planning to protect nature and human life
- Floods are responsible for over half of the people affected by extreme weather events and caused significant economic losses
- The atmosphere's capacity to hold moisture increases with increasing temperature indicating heavier rainfall events are more likely
- As the planet warms the risk of extreme rainfall and flooding is expected to increase
- In the United States there has been an increase in the number of intense rainfall events
- The devastating floods in Central Europe in June and September 2024 serve as a stark example of the impact of extreme rainfall

7.10 Longer and More Severe Droughts

- The extent of regions affected by droughts has also increased while evaporation has increased due to warmer conditions
- Higher temperatures also drive evaporation leading to drier conditions and prolonged droughts in many areas
- While natural weather systems can also play a role in droughts, climate change is making them more severe and longer lasting

- Drought, an unavoidable condition limits plant growth and agricultural productivity
- Plants have limited defense against drought and rely on various coping mechanisms to survive stress
- Under water deficiency conditions the two crucial ways to ensure plant growth are (a) drought escape and (b) drought tolerance
- Plant Growth Promoting Rhizobia (PGPR) enhances drought resilience by producing phytohormones such as auxins, gibberellins, and cytokinin's which regulate root growth and water uptake
- Arbuscular Mycorrhizal Fungi (AMF) support drought tolerance by improving nutrient and water absorption, strengthening root structures, and enhancing antioxidant defense systems that mitigate oxidative damage

7.11 Increased Storms

- Tropical storms and hurricane frequencies vary considerably from year to year, but evidence suggests substantial increases in intensity and duration since the 1970s
- In the extra-tropics variations in tracks and intensity of storms reflect variations in major features of the atmospheric circulation such as the North Atlantic Oscillation
- Studies using data from 75 weather stations of 15 countries in the Western Asia and Northern African region showed significant increases in the number of warm days and decreases in the number of cold days and revealed ample evidence of hot spots of vulnerability to climate change and extreme climate events
- With the increasing threat of global warming there is a drastic increase in the number of catastrophic storms and in the recent past storms caused the most significant economic losses and fatalities

7.12 Increased Wildfire Risk

- Climate change intensifies wildfires by creating hotter and drier conditions that fuel these events
- Increased evaporation due to higher temperatures dries out vegetation turning it into potential fuel
- Change in land use also influences wildfire risk and the overall threat of extreme wildfires is expected to rise in a warming world

7.13 Latest Reports of IPCC on Extreme Events

- The IPCC confirms that human-induced climate change is causing significant shifts in extreme weather events globally
- Climate change is significantly influencing the frequency and intensity of extreme events worldwide **(Fig:7.2)**

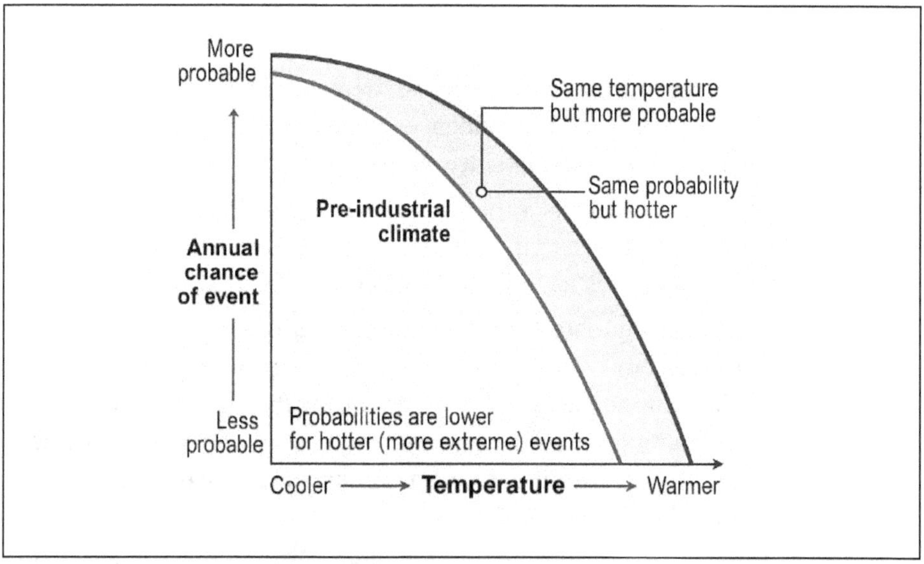

Fig. 7.2 Climate change and extreme events. Relationship between temperature and annual chance of event.

Source: IPCC,2021,2022, and 2023

- Preindustrial climate was cooler
- Current climate is warmer
- For the pre industrial climate the probabilities for annual chance of events are lower for hotter (more extreme) events
- However, under the current climate for same temperature there are more probabilities for annual chance of events
- Heatwaves: Heat waves virtually have increased in frequency and intensity globally since the mid-20th century due to human influence
- Heavy precipitation: Likely increased in frequency and intensity globally which is attributed to human-caused greenhouse gas emissions. This translates to increased risks of pluvial and riverine flooding
- Drought: Increasing trends in agricultural and ecological droughts observed in many regions, with human-induced increases in evapotranspiration playing a key role
- Tropical cyclones: Likely increase in the proportion of major intensity (category 3-5) and rapid intensification events over the past 40 years. Coastal hazards are exacerbated by rising sea levels and intense precipitation associated with tropical cyclones
- Compound events: Increased likelihood of combined events like concurrent heatwaves and droughts (leading to wildfire risk) and compound flooding (due to storm surge and heavy rainfall)
- These trends are projected to continue and intensify with further global warming, with even small increases in temperature leading to statistically significant changes in extremes
- Unprecedented extremes, both in magnitude and frequency, are expected to occur
- Specific regional impacts vary, with some regions experiencing stronger trends than others
- The number of weather-related natural catastrophes has risen on all continents since 1980 most notably in Asia and North America, which has had a dramatic impact on the insurance industry
- Human loss since the 1950s increased 50 percent each decade

7.14 Crop Damage and Kinetic Energy of Hailstones

- Kinetic energy is a significant factor in crop damage from hailstones, with higher energy correlating to increased damage potential. Several factors influence this relationship
- Hailstone size and velocity: Larger, faster hailstones possess greater kinetic energy, leading to more severe impact and damage. As "Hail Alert Technologies" explain, the impact energy of a 2-inch hailstone is over 22 times greater than that of a 1-inch hailstone. Wind can also increase hailstone velocity and impact energy
- Crop type and growth stage: Different crops exhibit varying susceptibility to hail damage. Example: The crop damage in case of peach is highest even at the lowest kinetic energy of hailstones. In contrast, the crop damage in case of corn(maize) is less even at highest kinetic energy of hailstones. However, literature indicate that corn is most vulnerable during the tasseling stage, while early-season damage can be less severe and overall corn damage due to hail energy is less. Crop damage is near moderate for moderate kinetic energy of hailstones in case of both vine and apple **(Fig:7.3)**

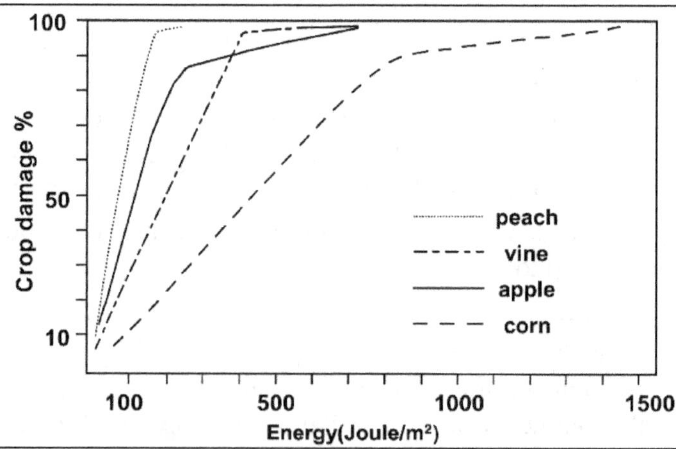

Fig. 7.3 Crop damage and kinetic energy of hailstones.

Source: Unknown. Class lecture notes collection

- Other factors: Hailstone hardness, shape, and accompanying winds further influence the severity of damage
- The kinetic energy of hailstones is a key indicator of potential crop damage, but it is important to consider the interplay of other factors to accurately assess the impact

7.15 Global Level Examples and Losses due to Extreme Events

- Hurricane Katrina which hit New Orleans in August 2005 caused an estimated 4000 fatalities
- The heat wave of August 2003 in western Europe resulted in approximately 35000 deaths. This was the hottest summer on record, and the analysis suggests that human influence has at least doubled the risk of a heat wave of this magnitude or greater
- In 2020, 389 natural disasters were reported killing 15,080 people, affecting 98.4 million others and costing 171.3 billion US$
- In 2021, Europe experienced an oppressive heat wave in summer and was hit by devastating floods in July. The magnitude of destruction was severe in Germany and Belgium causing several deaths
- From 1993 to 2022 extreme weather events resulted in nearly USD 4.2 trillion in direct losses
- Over the same period extreme weather events led to over 765,000 human loss globally
- In 2024 more than 824,000 people were displaced due to extreme weather events, the highest number of new displacements in 16 years

7.16 Impacts on Developing Countries and Small States

- Extreme events are responsible for many significant impacts in terms of both casualties and economic effects in developing countries and small states
- During the past few decades extreme events such as severe droughts, floods, storms and tropical cyclones, and wild land fires have caused major loss of human lives and livelihoods, the destruction of economic and social infrastructure as well as environmental damage
- Developed countries are better able to manage the impacts of climate change, but in developing countries and small states the increasing frequency and magnitude of extreme weather pose potentially disastrous consequences for agriculture and food security
- According to records the small states of the Pacific are disproportionately affected by extreme events with the average annual cost being much greater than larger countries

7.17 Mitigation and Adaptation Strategies

- Mitigation and adaptation to climate change involves taking actions to reduce vulnerability and build resilience to its impact. When considering climate elements like precipitation, drought, and rare floods, these actions can involve a range of strategies

7.17.1 Precipitation

- Changes in patterns: Climate change alters precipitation patterns, leading to more intense rainfall in some areas and increased drought in others
- Strategies
    - Green infrastructure: Utilizing approaches like bioretention (rain gardens) and permeable pavements to manage stormwater runoff and improve water quality

- Upgrading infrastructure: Redesigning culverts and drainage systems to handle increased rainfall volumes and prevent flooding
- Nature-based solutions: Restoring wetlands and floodplains to naturally absorb and filter floodwaters

7.17.2 Drought

- Impacts: Droughts reduce water availability, impacting agriculture, water supplies, transportation, and public health
- Strategies
    - Water conservation: Implementing measures like xeriscaping (using drought-tolerant plants), upgrading to water-efficient appliances, and capturing rainwater for reuse
    - Drought-resistant crops: Developing and planting crop varieties that can tolerate periods of low water availability
    - Efficient irrigation: Optimizing irrigation techniques to minimize water waste in agriculture
    - Water resource management: Monitoring surface water conditions and managing reservoirs and stream levels more effectively to optimize water use during droughts
    - Early warning systems: Utilizing drought prediction and early warning systems to enable proactive responses and planning

7.17.3 Rare Floods

- Increased intensity and frequency: Climate change is expected to increase the frequency and intensity of extreme weather events, including floods
- Strategies
    - Flood defenses: Constructing seawalls, floodgates, levees, and enhancing infrastructure to withstand higher flood levels
    - Relocation and zoning: Relocating infrastructure and buildings out of flood-prone areas and implementing zoning and land-use planning to limit development in vulnerable locations

  - Early warning systems: Developing and implementing early warning systems to provide timely alerts for impending floods, allowing for evacuation and preparedness
  - Nature-based solutions: Protecting and restoring mangrove forests as natural buffers against storm surges in coastal areas
  - Community engagement: Engaging local communities in planning and implementing flood adaptation strategies
- Adapting to climate hazards related to precipitation, drought, rare floods, etc., requires a multi-pronged approach involving infrastructural improvements, behavioral changes, ecosystem-based solutions, and collaborative planning at multiple levels

7.18 Addressing Climate Change Challenges of Sea Level Raising and Extreme Events

- The individuals, families, communities, governments, organizations, etc., have the best sciences available to plan for and implement climate-smart and resilient practices. Two critically important approaches which are available to managing climate resilience are mitigation and adaptation. These strategies are inter-related and must be pursued simultaneously as both can potentially improve resilience to the changing climate.

Chapter 8

Climate Change and Photosynthesis and Cultivated Soils

"The Earth is Mother; I am son of Earth". This excerpt recognizes Earth as a nurturing and life-sustaining entity, with humans as her children. "The Earth is held by Dharma in which auspiciousness gently pervades throughout the World.". This excerpt emphasizes that the Earth's existence and prosperity are rooted in cosmic law and order (dharma) and that this spiritual order brings blessings to the world. "She is Vishwambhara (all-bearing), She is Vasudhaa (producer of all Wealth), She is Pratishtha (foundation on which we live)." This excerpt indicates that Earth is the universal provider, the stable ground of existence for all living beings, and a source of sustenance and foundation for life itself "

8.1 Introduction to Photosynthesis

- Like all other organisms, the green plants use sugar and other oxidizable organic molecules as their source of energy
- However, unlike other organisms the green plants are autotrophic (self-feeding)
- Plants make their own food, chemically converting atmospheric CO_2 to sugar (carbohydrate) and related substances by means of radiant energy absorbed in the chloroplast which is a photosynthetic apparatus

- Plants are thus the primary producers in nature and are independent of any external supply of organic molecules
- In the process of photosynthesis, CO_2 and water are consumed and organic matter (sugar) and oxygen are formed, and radiant energy is stored as potential energy in the oxidizable organic molecules
- The oxygen liberated comes from the water consumed
- Thus, in this process the stable water molecule (H_2O) is cleared (photolyzed) by radiant energy absorbed in the chloroplast
- The resulting electrons are passed through intermediary carriers and used to reduce CO_2 to the level of carbohydrates
- Gas exchange in the leaf occurs through stomata while water is made available
- For most plants light is unlimiting in tropical areas but limiting in high latitudes depending on the season, but CO_2 becomes limiting during periods of bright Sun
- When plants are chlorotic (low in Chlorophyll) the pigment is a limiting factor
- The action spectrum for photosynthesis relating photosynthetic efficiency to wavelength of light shows that chlorophyll is the most effective pigment in photosynthesis, although other accessory pigments help by transferring their absorbed energy to chlorophyll
- In the overall process, termed photo phosphorylation, the ATP is synthesized when a photon gradient is established across the thylakoid membrane by the vectorial transfer of protons by the carrier plastoquinone and by protons from photolyzed water molecules
- When these photons diffuse out through special channels in the thylakoid membrane their energy is conserved through the synthesis of ATP from ADP and Pi (inorganic phosphate)
- Some of the electrons during water photolysis are terminated in a reduced carrier, NADPH (Reduced Nicotinamide Adenine Dinucleotide Phosphate)
- Together with ATP this compound reduces CO_2 to the level of carbohydrate

- ATP and NADPH are products of light reactions in photosynthesis; all rest are part of dark reaction
- The water molecule is effectively cleared
- After ionization, the hydroxyl ion transfers an electron to a light excited chlorophyll molecule

8.1.1 Photosynthetic Pathways

- There are 3 forms of photosynthetic pathways in plants viz., C_3, C_4, and CAM
- The crops following the "Calvin cycle" produce a 3-carbon compound as a precursor and are called C_3 plants
- Those crops following the "Hatch Slack pathway" produce a 4-carbon compound as a precursor and are called C_4 plants
- A very minor group of plants follow Crassulacean Acid Metabolism and are called CAM plants
- In all these three pathways, carbon must enter through the stomata into the plant
- C_3 plants like rice, wheat, etc., store the carbon obtained from the air in molecules with 3 carbon atoms and generally have intermediate rates of photosynthesis
- C_4 plants like maize, sorghum, etc., trap atmospheric carbon in 4 carbon molecules which are efficiently used in synthesizing primary sugars
- These plants have high rates of photosynthesis at optimum environmental conditions
- In CAM plants (dry climates) carbon is absorbed at night and is incorporated into carbon acids
- The gross photosynthesis of a leaf is equated to the flux of CO_2 into the leaf from the atmosphere
- Stomata must open to decrease resistance for photosynthesis in C_3 and C_4 plants
- In C_4 plants photosynthesis increases rapidly with PAR flux density
- There is no light saturation point for most of C_4 plants

- In C_3 plants with an increase in PAR flux density the photosynthesis is moderate. Light saturation point is reached
- The C_3 plants get light saturated at 2500-5000 foot candles
- Low CO_2 concentrations decrease the photosynthetic rates of both C_3 and C_4 plants
- The CO_2 compensation for C_3 plants is high and about 5-10 times greater than that of C_4 plants
- Therefore, photosynthesis continues to increase with increasing CO_2 levels in C_3 plants
- Examples
  - C_3 plants: Rice, Wheat, Soyabean, etc.
  - C_4 plants: Maize, Millets, Sorghum, Sugarcane, etc.
  - CAM plants: Members of Crassulaceae family. Dry climate crops
- The C_3 pathway is less efficient than the C4 pathway in production of photosynthates
- Since the C_4 plants have no light saturation, their photosynthesis is very little affected in the temperature range of 20-30 degrees centigrade
- Higher temperatures are required by C_4 plants compared to C_3 plants
- Thus, C_4 plants are more efficient under hot and bright weather conditions
- The C_3 plants have photorespiration i.e., high rates of respiration in PAR/ light conditions
- The C_4 plants have low photorespiration rates
- Therefore, the rate of photosynthesis markedly increases in C_4 plants with an increase in temperature provided other factors such as CO_2 and light are not limiting
- The temperature effects enzyme-controlled reactions in the dark stage

- When the intensity of light is low, the reaction is limited by the small quantities of reduced co-enzymes available so that temperature increases have little effect on the overall rate
- If the temperature is greater than 30 °C, the rate of photosynthesis abruptly falls due to thermal inactivation of enzymes

8.1.2 Photosynthesis and Respiration

- Respiration is burning (oxidation) of stored carbohydrates to produce energy
- Plant growth depends upon the excess of dry matter increase over the loss by respiration The net gain is known as "Net photosynthesis" or "Net assimilation rate"
- Photosynthesis takes place mainly in the leaves during the day
- However, respiration proceeds throughout the plant parts for the entire 24 hours
- In the process of respiration, plants burn up material to do work
- $CH_2O + O_2 \rightarrow CO_2 + H_2O$ + heat of combustion
- Respiration rates increase with temperature
- The respiration rate of leaves is usually 5 to 10 per cent of the gross photosynthesis at saturation light intensity and normal air
- At extremely low light intensities (forest floor) the net photosynthesis may be very small or even negative
- The point at which the respiration rate equals the photosynthetic rate is known as "Compensation point" which is a function of light intensity and temperature
- Day and night temperatures exert opposite effects on net photosynthesis
- A high night temperature increases the respiration loss and so reduces the net photosynthesis
- Day temperatures up to 30 °C may increase net photosynthesis
- In general, the net photosynthesis of cereals is positively correlated with daily maximum temperature but negatively correlated with night temperature

- Since the ideal condition for net photosynthesis is a combination of relatively high daytime temperatures and low nighttime temperatures, a large diurnal temperature range is often desirable
- Plant leaves infected with virus have a lower photosynthesis
- Photosynthetic rates of plant leaves usually decline with age
- Net assimilation rates of top, middle, and bottom leaves differ a lot
- As the leaf age decreases from the bottom to the top of the plants, the net assimilation rates increase in that order
- It has been estimated that even the most efficient plants are able to synthesize the energy of only 5 per cent of PAR, which is known as Radiation Use Efficiency (RUE)
- Energy which is not directly used for photosynthesis is mainly accounted for by the heating of plant tissue, with some of the radiant energy being lost by transpiration, shortwave reflection, and re-radiation in the long waves

8.2 Effect of Temperature on Photosynthesis and Respiration Ratio of Tomato

- High temperatures negatively affect the balance between photosynthesis and respiration in tomato plants **(Fig:8.1)**
- Up to 20 degrees centigrade low respiration, maximum dry weight production, and slow development takes place
- Between 20 and 35 degrees centigrade high respiration, low dry weight production, and rapid development takes place
- Tomato plants thrive and maximum photosynthesis and overall growth occurs at an optimum temperature range of 20 to 35 degrees centigrade
- High temperature effects: When temperatures rise above optimal levels (above 35°C), several detrimental effects occur in tomato plants
- Photosynthesis Inhibition in tomato

- High temperatures disrupt the chloroplast, where photosynthesis takes place
- They inhibit chlorophyll formation and activity, leading to decreased photosynthetic rates

- Photosystem II, crucial for light energy conversion, becomes vulnerable to damage at high temperatures

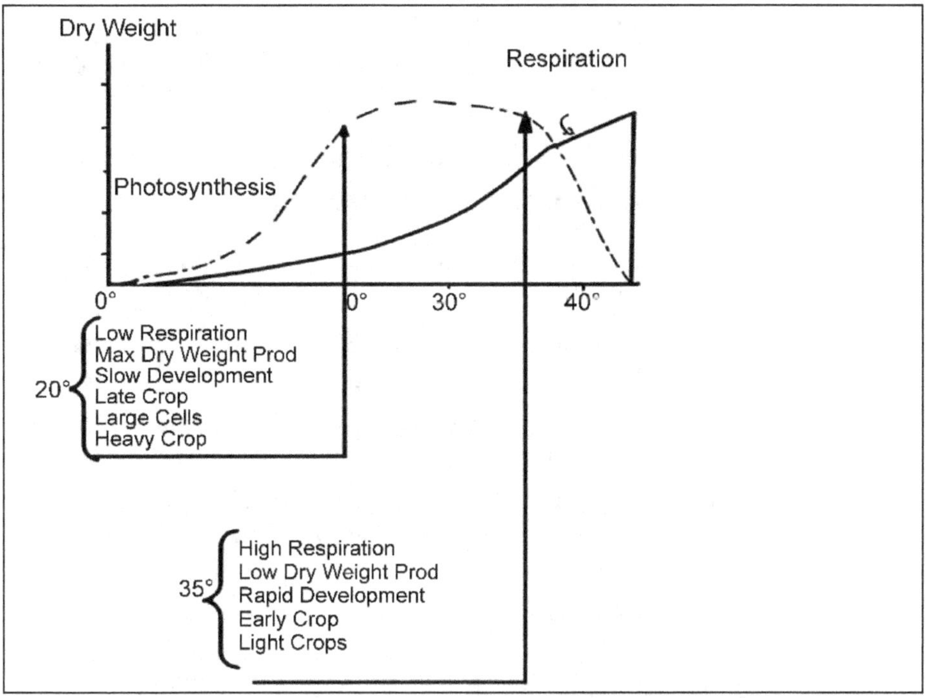

Fig. 8.1 Effect of temperature on photosynthesis and respiration of tomato.

Source: Unknown. Class lecture notes collection

- Increased Respiration in tomato
  - Respiration rates generally increase with temperature, burning off the sugars produced by photosynthesis at a faster rate
  - At extreme temperatures above 35°C, the rate of respiration exceed the rate of photosynthesis, leading to a net loss of energy for the plant, hindering growth and productivity

- Reduced reproductive performance in tomato
    - o High temperatures negatively impact flowering, pollination, fruit set, and fruit development, further reducing the overall yield of tomato
    - o High temperatures shift the balance towards increased respiration and decreased photosynthesis, ultimately diminishing the tomato crop ability to produce the sugars it needs for growth and fruit development. This results in lower yields, poor fruit quality, and increased susceptibility to other stressors

8.3 Generalized Relationship between Photosynthesis and Light Intensity

- Light is essential for photosynthesis, and its rate increases proportionally with light intensity up to certain levels
- It is because more chlorophyll molecules absorb light energy to power the light-dependent reactions
- However, this increase eventually plateaus at a point known as the "Light saturation point"
- Beyond this point, further increases in light intensity will not significantly boost the rate of photosynthesis
- At very high light intensities, the rate might even decrease due to potential damage to chlorophyll molecules (photoinhibition)
- Under adequate soil moisture as light intensity increases optimum possible photosynthesis takes place **(Fig:8.2)**

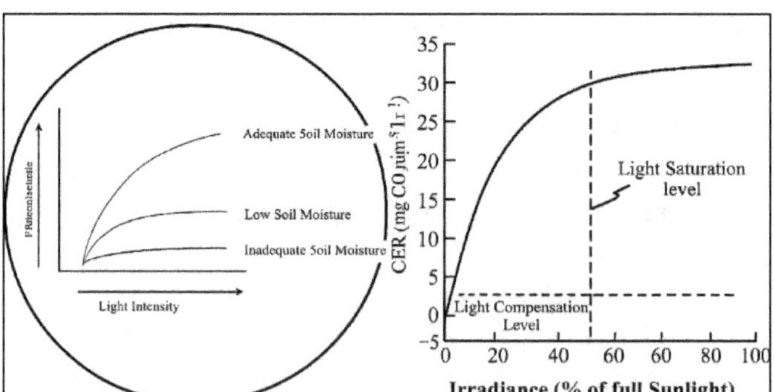

Fig. 8.2 Generalized relationship between photosynthesis and light intensity. **Source:** Unknown. Class lecture notes collection

- Under low soil moisture, as the light intensity increases the photosynthesis will be less than moderate
- Under inadequate soil moisture the photosynthesis will be much lesser than moderate

8.4 Generalized Relationship between PAR/Light and Net Carbon Dioxide Uptake

- Light Saturation point
  - The light saturation point is the specific intensity of light at which a plant's photosynthetic rate reaches its maximum, and any further increase in light intensity will not increase the rate of photosynthesis. This plateau occurs because other factors, such as CO_2 availability or enzyme capacity become limiting, preventing the plant from using the extra light to produce more energy and develop. Different plant species have different light saturation points, with shade-tolerant plants typically having lower saturation points than sun-loving plants **(Fig:8.3)**

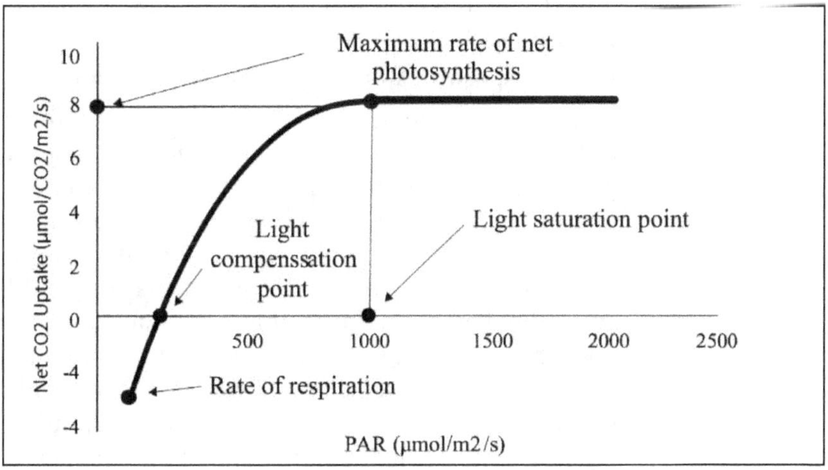

Fig. 8.3 Generalized relationship between PAR/Light and net carbon dioxide uptake.

Source: Unknown. Class lecture notes collection

- Light Compensation point
  - The light compensation point (LCP) is the specific level of light intensity where a plant's photosynthetic rate (CO_2 absorbed and oxygen released) exactly equals its respiration rate (CO_2 released and oxygen absorbed). At this point, the plant isn't growing or shrinking, but simply maintaining itself. For growth to occur, the light intensity must be above the LCP
- Maximum rate of net photosynthesis
  - The maximum rate of net photosynthesis varies widely depending on the plant species and environmental conditions. This maximum rate is also known as "Pn-max" and is achieved under conditions of saturated photosynthetic active radiation, optimal temperature, and sufficient CO_2 concentration
  - Photosynthesis and PAR: The PAR encompasses the wavelengths of light (400-700 nanometers) that chlorophyll and other pigments in plants absorb to drive photosynthesis. The rate of photosynthesis, and thus CO_2 uptake, is directly influenced by the availability and intensity of PAR
  - PAR intensity and CO_2 Uptake: As PAR intensity increases, the rate of photosynthesis, and therefore the rate of CO_2 uptake, generally increases, up to "Light saturation point". The excessively high PAR can lead to photoinhibition, a phenomenon where photosynthetic rates can decrease due to light stress. The relationship between PAR and net CO_2 uptake is fundamental to plant life and growth. Understanding this relationship, including concepts like light saturation and compensation points, as well as the effects of different PAR wavelengths, allows for optimizing light environments to maximize plant productivity, especially in controlled environments
- Impact of climate change on PAR/Light and CO_2
  - CO_2 fertilization effect: Increased atmospheric CO_2 concentrations can initially stimulate photosynthesis and enhance plant growth leading to a greater uptake of CO_2

- Climate-mediated impacts: However, climate change also introduces other factors that can counteract or reduce this initial benefit
    - Increased temperatures in some cases negatively impact photosynthetic efficiency
    - Altered precipitation patterns and drought increases in vapor pressure deficit (drier air) due to warming can cause water stress and reduce photosynthesis. This can weaken the indirect effect of CO_2 on vegetation carbon uptake
    - Changes in cloud cover and irradiance can limit photosynthesis and carbon uptake, particularly in tropical forests. Diffuse radiation, however, can sometimes cause less photosynthetic saturation of vegetation

- While elevated CO_2 can initially boost plant CO_2 uptake, climate change and its associated effects like altered temperatures, water availability, and radiation patterns introduce complexities that can ultimately reduce the long-term benefit of CO_2 fertilization

8.5 Relationships among Soil and Atmospheric Moisture Stresses, Photosynthesis (P) and Light Intensities (I)

- As light intensity increases the photosynthesis reduces abnormally under high soil moisture stress
- As light intensity increases moderate photosynthesis occurs under medium soil moisture stress
- As light intensity increases high photosynthesis occurs under low moisture stress
- Under extreme stress of atmospheric conditions, the photosynthesis abruptly decreases as light intensity increases
- Under high stress atmospheric conditions, as light intensity increases moderate photosynthesis continues **(Fig: 8.4).** The reasons are as follows

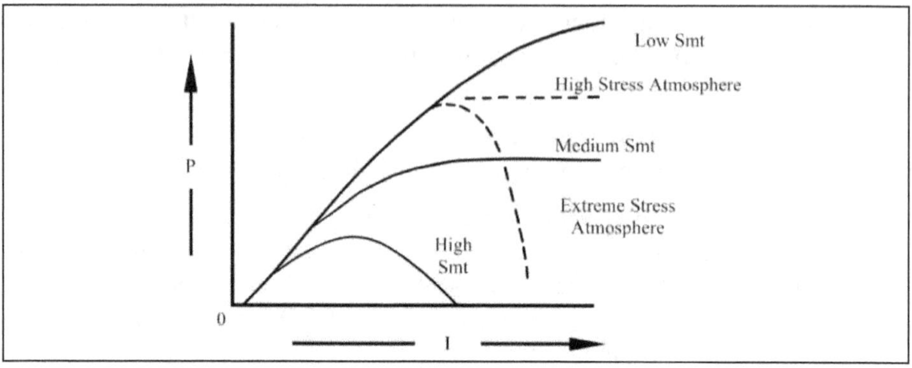

Fig. 8.4 Relationships among soil and atmospheric moisture stresses, photosynthesis (P) and light intensities (I).

Source: Unknown. Class lecture notes collection

- Relationships among SMS and AMS and Photosynthesis (P)
  - Soil moisture stress (drought): Soil moisture stress reduces photosynthesis due to stomatal closure (limiting CO_2 intake), chlorophyll degradation, enzyme inhibition, and damage to photosynthetic structures. The effects can be categorized as stomatal limitation (mild to moderate stress) and non-stomatal limitation (severe stress)
  - Atmospheric moisture stress (high VPD/low humidity): The high VPD also limits photosynthesis by driving stomatal closure to conserve water. High VPD can induce stomatal closure even with ample soil moisture, thus reducing photosynthetic rates
  - Excessive soil/atmospheric moisture (waterlogging/high humidity): While less commonly studied than drought, excessive soil moisture can also negatively impact photosynthesis by impairing root respiration, nutrient absorption, and gas exchange
- Interaction of SMS with light intensity (I)
  - Drought and light intensity: The negative impact of drought on photosynthesis can worsen under high light intensity. High light, coupled with water stress, can cause photoinhibition, where excess light energy damages the photosynthetic machinery due to a lack of CO_2. Plants may adapt by using weak light more efficiently under drought conditions

- Fluctuating light: Drought stress can delay the photosynthetic induction period in fluctuating light, reducing carbon assimilation. It can also accelerate photoinhibition of Photosystem- I (PS-I) under such conditions
- Plants try to balance carbon dioxide uptake and water loss through stomatal regulation. Both insufficient (drought/low humidity) and, to a lesser extent, excessive (waterlogging/ high humidity) soil and atmospheric moisture can significantly stress plants. This impacts their photosynthetic capacity, especially with high or fluctuating light intensities

8.6 Summary of Interactions and Complexities among Weather Elements and Photosynthates

- Carbon dioxide
  - Elevated CO_2 helps crops to cope with drought conditions better
  - C_4 crops (maize, sorghum, pearl millet) require less water per gram of carbon assimilated than C_3 crops (rice, wheat)
  - Therefore, sorghum and maize are preferred in dry land areas
  - Research results indicate that on average a doubling of CO_2 will increase the yields of many crops by 30 per cent
  - The Carbon dioxide compensation for C_3 plants is high and about 5-10 times greater than that of C_4 plants
  - For a doubling of CO_2 an increase in yield varies amongst crops being (a) 9 per cent for maize (b) 15 per cent and 28 per cent respectively for rice and wheat and 50 per cent for soybean
  - Elevated CO_2 increases photosynthetic rates and plant growth of natural ecosystems and of fruit trees like orange, citrus, apple, etc.
  - It is reported that sour oranges exposed to CO_2 concentrations of 300 ppm over the normal produced 3 times more biomass than control (ambient carbon dioxide) in the first 2 years after which the rate of increase in biomass began to decline to a steady value of 1.7 times the control from the 9th year

- Solar Radiation
    - Maize, tobacco and soybean are the most efficient, least efficient, and average users of PAR respectively
    - Reduction of bright sunshine hours during specified crop stages of growth of several crops reduces crop yields
    - Reduced solar radiation in the vegetative phase of rice has little influence on yield except under excessively cloudy conditions
- However, solar radiation from panicle initiation to maturity is linearly related to the yield of rice. Research results indicate that reduction in sunlight in the six (6) week period following panicle initiation in rice reduces grain number, increase spikelet sterility, and ill filled spikelets and hence lower rice yields
- In mixed cropping system, the taller crop should be of the C_4 type and the shorter one of the C_3 type to achieve optimum crop yields. Example: (a) Maize (tall crop) intercrop with Soybean (short crop) is beneficial
- Wind
    - Wind effects plant growth in three significant ways viz., transpiration, carbon dioxide intake, and mechanical breakage of leaves and branches
    - Changes in wind speed indirectly affects crop surfaces by changing the resistance, thus controlling the fluxes of heat, water vapor, carbon dioxide, and pollination
    - At high altitudes and low atmospheric pressures, the solubility of carbon dioxide and oxygen in water are reduced
    - Therefore, some plants show stunted growth at higher altitudes as concentrations of oxygen and carbon dioxide reach low concentrations
- Rainfall
    - Water is a reagent in photosynthesis
    - Crop water use is a dynamic process governed by a complex interplay of environmental, crop, and management factors. Optimizing water management during the growing season is critical for both yield and water conservation

- WUE is an important consideration in the context of climate change, as it can help identify crops that are more resilient to water scarcity

- Temperature
    - When the temperature increases to three (3) degrees centigrade above normal, the beneficial effect of doubled CO_2 gets canceled out for any crop in any season and any location
    - A doubling of CO_2 leads to an increase of temperature from 3 to 5 degrees centigrade. Thus, in real time an increase in elevated CO_2 leads to a decline in crop yields
    - Increased temperature also increases the photorespiration of C_3 crops. Thus, the effects of increased temperature on yield reduction will be more for C_3 than C_4 crops
    - Currently temperatures in India are one degree centigrade above optimal for rice and wheat. Therefore, further global warming adversely affects rice and wheat crop yields necessitating technological interventions
- Photosynthetic process, C_3 and C_4 plants
    - In C_4 plants photosynthesis increases rapidly with PAR flux density. Example: Maize is the most efficient crop of PAR
    - There is no light saturation point for most of C_4 plants
    - In C_3 plants with an increase in PAR flux density the photosynthesis is moderate. Example: Soybean
    - The C_3 plants get light saturated at 2500-5000 foot candles
    - Low CO_2 concentrations decrease the photosynthetic rates of both C_3 and C_4 plants
    - Photosynthesis continues to increase with increasing CO_2 levels in C_3 plants
    - Examples
        - C_3 plants: Rice, Wheat, Soyabean, etc.
        - C_4 plants: Maize, Millets, Sorghum, Sugarcane, etc.
        - CAM plants: Members of Crassulaceae family. Dry climate crops

- The C_3 pathway is less efficient than the C_4 pathway in production of photosynthates
- Photosynthesis of C_4 plants is very little affected in the temperature range of 20-30 degrees centigrade
- Higher temperatures are required by C_4 plants compared to C_3 plants
- Thus, C_4 plants are more efficient under hot and bright weather conditions. Example: The effects of increased temperature on yield reduction will be more for C_3 than C_4 crops
- The C_3 plants have photorespiration. High rates of respiration in light conditions
- The C_4 plants have low photorespiration rates
- In general, the rate of photosynthesis markedly increases with an increase in temperature provided other factors such as CO_2 and light are not limiting

- Summary
  - The combined effects of weather elements on photosynthesis are complex. Example: Increased temperature often negates some of the benefits of increased CO_2
  - C_3 and C_4 plants respond differently to elevated CO_2 and temperature. Example: C4 plants like corn (maize) typically show less benefit from increased CO_2 levels
  - While CO_2 fertilization may increase yields, it can also lead to lower nutritional content, creating a trade-off between quantity and quality
  - Climate change impacts on photosynthesis vary significantly by region, with some areas experiencing more severe effects than others due to variations in weather elements and other factors

8.7 Generalized Impacts of Climate Change on Photosynthesis

- The impact of climate change on photosynthesis is complex and varies depending on factors like plant species, location, and specific climate change impacts (warming, changes in precipitation, etc.)
- Climate change impacts photosynthesis both positively and negatively. Increased carbon dioxide levels can enhance photosynthesis in some plants, a phenomenon known as "Carbon fertilization"
- However, rising temperatures and altered precipitation patterns due to climate change can negatively affect photosynthetic efficiency, potentially reducing plant growth and crop yields
- Water availability, another factor influenced by climate change, also plays a crucial role in photosynthesis, affecting nutrient uptake and overall crop growth and development
- From photosynthesis point of view, globally, the yields of C_3 crops (rice, wheat, vegetables, grapes, etc.) generally increase. Yields of C_4 crops (maize, sorghum, pearl millet, sugarcane, tropical grasses, etc.) generally decrease. However, this depends on several other factors
- Positive Impacts
    - Fertilization: Higher concentrations can enhance photosynthesis in many plants leading to increased biomass production and potentially higher yields in crops
    - Increased Water Use Efficiency: Plants may become more efficient at using water under elevated CO_2 conditions, potentially reducing water stress in some regions
    - Warm Temperature Adaptation: Some crops, particularly those in warmer regions, may show increased photosynthesis rates at slightly higher temperatures, but this effect is limited
- Negative Impacts
    - While some crops may exhibit increased yields under higher CO_2, the nutritional value, particularly protein and micronutrient content can decline

  - Changes in precipitation patterns, leading to drought can negatively impact nutrient uptake, root development, and overall plant health thus affecting photosynthesis
  - Heavy rainfall can lead to waterlogging which reduces oxygen availability for roots and can inhibit nutrient uptake, ultimately affecting photosynthesis
  - Climate change can affect stomatal conductance, which regulates CO_2 uptake and water loss

8.8 Mitigation and Adaptation Strategies for Photosynthesis under Climate Change Scenarios

- The following mitigation and adaptation strategies are essential
  - Developing crops with better heat and drought tolerance
  - Adjusting planting times/ sowing dates
  - Optimizing irrigation and fertilization and their application schedules etc.

8.9 Cultivated Soils

8.9.1 Introduction to Micrometeorology of Cultivated Soils

- Energy exchanges and transport take place at the surface of the soil which influence the climate near the ground because of changes in moisture status and temperature of upper soil layers
- Ploughing the soil, mulching, shading, wind breaks, etc., modify the microclimate of the cultivated soils
- Some of the radiation striking the surface will be conducted into the ground
- Heat energy is thus stored in the ground during the day, only to be released from the ground into surface at night. Therefore, cultivated soil acts as a "Sink" of heat during the day and "Source" of heat at night
- Small-scale atmospheric conditions in the immediate vicinity of the soil and crops play a critical role in agricultural productivity and sustainability

- The dynamic exchange of energy, momentum, and mass between the land surface and the atmosphere includes interaction among "Solar Radiation, Wind (air), Temperature (thermal energy /heat), and Water"
- Intricate relationships exist among agricultural practices, soil properties, plant growth, and the small-scale atmospheric environment, which ultimately contribute to a more sustainable and productive agricultural system
- In addition, interactions among the near-surface atmosphere, agricultural crops, and the soil environment are crucial for understanding and optimizing agricultural practices which lead to improved crop yields and sustainable land management
- Also, micrometeorological conditions affect all crop development stages viz., germination, flowering, seed setting, and final crop yield
- Turbulent transport and energy exchange between the atmosphere and the biologically active canopy takes place
- In summary, the micrometeorology of cultivated soils highlight the vital connections among weather, plants, and soil in agricultural systems. By understanding and leveraging these interactions, it is possible to improve crop production, optimize resource use, and promote more sustainable farming practices under climate change scenarios

8.9.2 Soil Moisture and Microclimate

- The amount of water in the soil significantly impacts the microclimate and ultimately crop growth and yield
- Moist soils buffer temperature extremes creating a more favorable environment for plant growth
- Evaporation of soil moisture contributes to cooling the surrounding air, particularly in conditions with high topographic heat load and higher soil moisture levels
- Canopy density and topographic features (aspect and slope, position, etc.) interact with soil moisture to modify the microclimate, affecting temperature and evapotranspiration rates

8.9.3 Crop Canopy and Soil Microclimate

- The architecture of the crop canopy, including leaf area index (LAI) and leaf inclination, influences how radiation is intercepted, absorbed, and reflected
- Crop canopies influence local air temperature and humidity creating a microclimate different from the surrounding environment
- Wind patterns within and above the canopy affect gas exchange (CO_2 and water vapor) and the distribution of heat and moisture

8.9.4 Soil Microclimate and Improved Agricultural Practices

- By monitoring soil moisture and evapotranspiration rates, farmers can schedule irrigation efficiently, minimizing water waste and ensuring adequate soil moisture for crops
- Managing microclimates within crop canopies can create optimal conditions for growth leading to improved crop quality and higher yields
- Micrometeorological studies help understand how crops respond to changing weather patterns, aiding in developing strategies to adapt to climate change, and mitigate its effects
- Understanding the microclimatic conditions that favor pests and diseases can help farmers implement targeted control measures
- In cultivated soils turbulent exchanges of heat, momentum, and mass (including water vapor and gases like CO_2), and momentum takes place between the surface and the atmosphere
- The properties of the soil surface layer, including color, water content, texture, and density affect how incident radiation is partitioned among evaporating water, warming the air, and warming the soil itself
- These studies help in understanding ecosystem functioning, agricultural productivity, and climate change continued

advancements in data acquisition, modeling, and modification techniques are crucial for addressing the challenges posed by a changing climate and for promoting sustainable land management practices

- To effectively solve the climate change issues and problems, micrometeorological principles and technics involving solar radiation, wind, temperature, and water that play among themselves in cultivated soils must be applied comprehensively

8.10 Addressing Climate Change Challenges of Photosynthesis and Cultivated Soils

- The individuals, families, communities, governments, organizations, etc., have the best sciences available to plan for and implement climate-smart and resilient practices. Two critically important approaches which are available to managing climate resilience are mitigation and adaptation. These strategies are inter-related and must be pursued simultaneously as both can potentially improve resilience to the changing climate.

Chapter 9

General options for Climate Change, Climate Neutral, and Net Zero Emissions

"Humans are seen as integral to nature with a responsibility to nurture Earth as a mother, ensuring prosperity for all life. The interconnectedness of all existence is revered guiding actions to preserve the sacred balance for future generations. May the Earth be our mother and the heavens our father; may they protect and purify us, fostering harmony and well-being"

9.1 Introduction to General Options for Climate Change

- Climate change "Mitigation" is trying to "Reduce" the expected climate change impact. Examples: New policies, innovative technologies, changes in day-to-day activities, etc.
- Climate change "Adaptation" is preparing to "Cope with" inevitable changes ahead due to climate change. Examples: To cope with higher temperatures, scarcer water resources, more frequent storms etc., occurring at present or anticipating such changes in future. An agroecosystem with diversified crops has greater potential for adaptation to climate change
- Coal and oil paved the way for the industrial progress of the developed world

- The developing countries are also taking the same path in search of equal living standards
- The dependence on carbon-based energy has caused a significant buildup of greenhouse gases in the atmosphere
- The IPCC reported that global warming due to climate change is happening, and it is unequivocal
- It also mentioned that the cause is carbon dioxide and other greenhouse gases
- It is not only burning carbon but also valuable forest trees being felled for timber and making paper
- Trees are also being felled to supply growing demand for biofuels
- This further manifestation of greenhouse gas habit is releasing vast amounts of carbon dioxide and destroying a valuable resource for absorbing atmospheric carbon dioxide, which is further contributing to climate change
- The environmental, economic, and political implications of global warming are profound
- Ecosystems, from mountains to oceans, from the poles to the tropics are undergoing rapid change
- Low- lying cities are facing inundation, fertile lands are turning into desert, and weather patterns are becoming ever more unpredictable
- Everyone is responsible for this, and the cost must be borne by all
- The poor will be the hardest hit by weather related disasters due to soaring price inflation for staple foods, and even the richest nations face the prospect of economic recession and a world in conflict over diminishing resources
- Mitigating climate change, eradicating poverty, and promoting economic and political stability all demand the same solution "The world must not immerse in the carbon habit"

- From reducing consumption and increasing energy efficiency to offsetting emissions via the multitude of carbon trading schemes, the opportunities are plentiful
- For everyone the scientists, individuals, a business, an organization, a government, etc., there are many steps to reduce climate footprint

9.2 General options for Climate Change "Mitigation"

- Warming caused by human activities could lead to "Abrupt or irreversible" impacts
- Climate change is not a smooth linear process of a world warming gradually and steadily, but rather a series of sudden jolts, like flips from one stable climate to another, radically different
- The pace of change is already much faster than expected
- Without drastic change its impacts appear certain
- The key to success for an effective emissions "Reduction" program is to have a well-organized, performing structure, and a clear process in place
- The following "All in one" strategies are sorted by cost efficiency for mitigation

9.2.1 Savings

- Standby losses
- Use of sugarcane biofuels
- Fuel efficient commercial and other vehicles
- Water heating
- Air conditioning
- Lighting systems
- Fuel efficient commercial vehicles

9.2.2 Costs

- Nuclear
- Small transit
- Small hydro
- Industrial non- CO_2
- Airplane efficiency
- Forestation
- Coal to gas shift
- Biodiesel

9.3 General options for Climate Change "Adaptation"

- Adaptation actions are taken to "Cope with" a changing climate. Adaptation aims at coping with risk and damage from current and future harmful impacts cost effectively for exploiting potential benefits
- Examples of options for climate change "Adaptation"
    - Using scarce water more efficiently
    - Adapting existing building codes to withstand future climate conditions and extreme weather events
    - Construction of flood walls and raising levels of dykes against sea level rise
    - Development of drought tolerant crops
    - Selection of forestry species and practices less vulnerable to storms and fires
    - Development of special plans and corridors to help species migrate etc.

9.4 General options for Climate Change as Individuals

- Individual responsibility for climate change mitigation and adaptation decreases with decreasing economic power

- In poor countries more responsibility lies with governments and companies which actually act
- As a consumer
  - Buy high quality
  - Choose seasonal products
  - Choose local products
  - Try organic products
  - Drink tap water
  - Choose products with limited packaging
- As a resident
  - Turn off electric devices when not using them (make sure that they do not remain in stand-by mode)
  - Turn off the light when leaving a room
  - Collect the rainwater for gardening
  - Choose low energy bulbs
  - Put a lid on pans when boiling water
  - Run washing machines during slack hours
  - Improve insulation (window, roof walls, etc.)
  - Replace very old electric device
  - Use water saving tap inserts
  - Use water saving shower heads
  - Choose collective instead of individual building
  - Choose ecological material, locally extracted and manufactured
  - Choose renewable energies
- As a mover
  - Limit flying
  - Limit car use
  - Replace very old cars
  - Limit speed
  - Drive with fluidity
  - Respect pedestrians and bikers
  - Use car only if there is no other option
  - Bike or walk

- Use video conferencing instead of travel
- Use public transportation

- As a citizen
  - Spread the word of climate change
  - Get involved
- As a parent
  - Educating children to save energy
  - Educating children to use resources when needed
- As a professional
  - Invest in renewable energies
  - Invest in low energy sectors
  - Turn off computers when leaving
  - Switch off printers and copy machines at night (make sure they do not remain in stand-by mode)
  - Print only when necessary
- At national/international scale
  - Set performance standards of power plants, gas processing plants, refineries, chemical industries, cement factories, and iron and steel manufacturing
  - For transport, expand the national public transport and develop rail and river freight alternatives
  - In agriculture, limit/control fertilizers and pesticides, preserve biological sinks like forests, follow organic farming wherever possible
  - In energy- prefer solar, wind, and ocean
  - Organize awareness campaigns and educational campaigns
  - In construction promote local and ecological building and set public buildings as an example
  - In waste management, support eco design projects, organize sorting, and recycling of waste
  - In urban planning limit urban sprawl, control, and limit the use of cars in city limits
  - In public transport, expand the public transport network, run a reliable and regular service, make it affordable, make it easy for everyone to use

9.5 Tackling Climate Change

- Climate change mitigation and climate change adaptation complement each other and together significantly reduce the consequences of climate change
- Greenhouse gases have both natural and manmade sources
- There are many natural processes that release and store GHGs like volcanic activity and swamps which account for considerable amounts of GHG emissions
- Their concentration in the atmosphere also varied in pre-industrial times
- But at present atmospheric concentrations of CO_2 and CH_4 far exceed the natural range 6,50,000 years ago
- These enormous amounts of GHGs are closely linked to human activities, such as fossil fuel combustion and land use change that release GHGs into the atmosphere
- Nature is not capable of balancing this development
- Individual responsibility for climate change mitigation and adaptation decreases with decreasing economic power
- In order to stay below global 1.5 degrees centigrade temperature rise, the UN Development programs suggested emissions reductions by developed countries
- The developing countries are also suggested to cut their emissions slowly, but steadily such that both developed and developing countries would converge between 2060 and 2070 in order to stay below 1.5 degrees centigrade temperature rise

9.6 Adaptation to Climate Variability in Indigenous Civilizations

- Throughout history, human societies have adapted to climate variability. Example: Adaptation amongst pre-Colombian communities of Latin America. For pre-Colombian indigenous civilizations in the high lands of present-day Latin America, one of the most critical limitations affecting development was the irregular distribution of water. In large areas, human activities depended on seasonal rainfall, and a range of adaptive actions were developed to satisfy requirements. Engineering activities included rainwater cropping, filtration and storage; the construction of surface and underground irrigation channels, including devices to measure the quantity of water stored, rectification of river courses and the building of bridges. Water was also used to cut stone blocks for construction: stones were cut in regular geometric shapes by leaking water into cleverly made interstices and freezing it during Altiplano nights which reach below zero temperatures. These efforts enabled the subsistence of communities which, at the peak of the Inca civilization, included some 10 million people in what Peru and Ecuador is today

9.7 Adaptation to Present Day Climate Change by Modern Civilizations

- There are innumerable examples of adaptation to present day climate change by modern civilizations. A few among these are
- Case study of Perth, Australia water supply
- Tsho Rolpa Risk Reduction Project in Nepal
- Adaptation to the rising sea level, The Maeslantkering in the Netherlands

9.7.1 Case Study of the Perth, Australia Water Supply

- In Perth, Australia, the shortfall of water has been met through the construction of two dams, additional ground water extraction, and construction of a desalination plant which supplies 140000 cubic meters per day with design for expansion to 250000 cubic meters per day

9.7.2 Tsho Rolpa Risk Reduction Project in Nepal

- The Tsho Rolpa is a glacial lake located at an altitude of about 4580 meters in Nepal. Glacial shrinkage increased the size of the Tsho Rolpa from 0.23 square km in 1957/58 to 1.65 square km in 1997. The 90-100 million meter cube of water, which the lake contained by this time, were only held by a moraine dam – a hazard that called for urgent action to reduce the risk of a catastrophic glacial lake outburst flood. The Government of Nepal with the support of international donors, initiated a project in 1998 to lower the level of the lake by 3 meters by cutting a channel in the moraine. Meanwhile, an early warning system was established in 19 villages downstream in case a Tsho Rolpa GLOF should occur despite these efforts. In 2002, the four-year construction project was completed at a cost of about US$ 3.2 million

9.7.3 Adaptation to the Rising Sea Level, The Maeslantkering in the Netherlands

- The Maeslantkering in the Netherlands is the largest moving structure on Earth. It protects the city of Rotterdam and the surrounding towns and agricultural areas against flooding from the sea. Completed in 1997, it is designed to close when a storm surge of 3 meters above normal sea level is anticipated in Rotterdam. Under present day conditions this should be once every ten years. In 50 years of time due to sea level rise this rate of closure is expected to increase to once every 5 years. The 13 km long Confederation bridge connects Prince Edward Island and was completed in 1997. It is designed to withstand a 1 meter rise in sea level during its 100 year design lifetime

9.8 Natural – Capital Framework for Changing Climate

- A "Natural - Capital" framework recognizes that the planet's ecosystems and the services they provide (clean air and water, climate regulation, biodiversity) are essential assets and foundational to human well-being and economic activity
- It elevates "Nature" as an integral component in addressing climate change, recognizing the intricate connections between climate and biodiversity
- Community, educators, scientists, and policy/ public come under "Natural" component of frame work. The "Natural" component also takes into account environment and economy; forest issues; ecosystems and species/ genetics

- Conservation and management; education and awareness; science and technology; and sustainable use and value come under "Capital" part of frame work. The "Capital" part also takes into account issues at global- national; national- regional; regional- local and local- site levels **(Fig:9.1)**

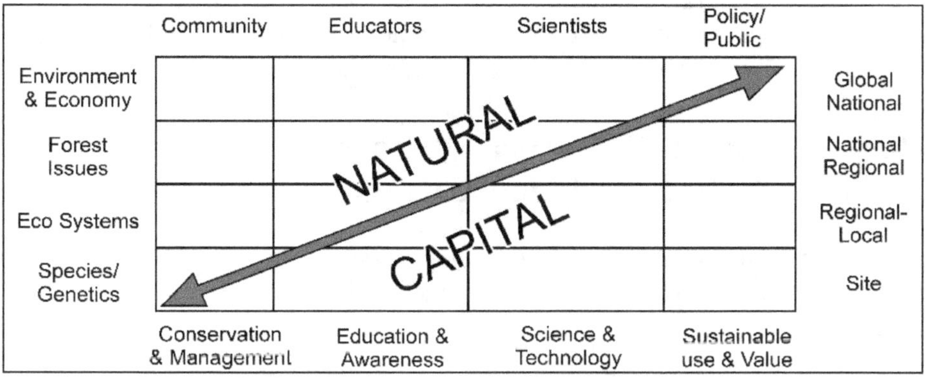

Fig. 9.1 Natural – Capital framework for changing climate.
Source: Unknown. Class lecture notes collection

- Understanding dependencies and impacts: It helps identify how organizations and economies rely on "Natural – Capital" (like forests for carbon sequestration and coastal ecosystems for storm protection) and how climate change impacts these assets and their services
- Integrating "Natural – Capital" into decision-making: The framework promotes incorporating the value of natural assets and ecosystem services into economic and policy decisions related to climate mitigation and adaptation, shifting the focus from simply considering financial and manufactured capital. This highlights that "Natural-Capital" is a missing piece in climate policy
- Promoting nature-based solutions: It encourages investing in the restoration and protection of natural ecosystems as a cost-effective and beneficial means of addressing climate change and building resilience. Examples: Reforestation, wetlands restoration, agroforestry, etc.

- Enhancing climate resilience and adaptation: Healthy ecosystems contribute to climate resilience by regulating water flows, protecting against extreme weather events, and supporting adaptation strategies
- Sustainable economic development: By valuing and investing in "Natural – Capital", societies can work towards a more sustainable economic development path that acknowledges the long-term dependency on healthy ecosystems

9.9 Adaption Cycle through Space and Time to Climate Change

- Adaptation to climate change operates through a continuous cycle, evolving over both space and time **(Fig:9.2)**

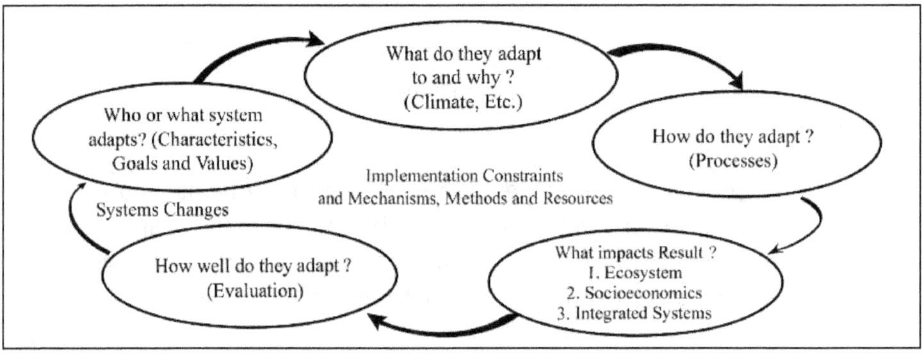

Fig. 9.2 Adaption cycle through space and time to climate change.
Source: Unknown. Class lecture notes collection

- This cycle involves
  - System changes: This part of cycle moves between Who or what system adapts? (characteristics, goals and values) and how well do they adapt? (evaluation)
  - Implementation constraints and mechanisms methods and resources: This part of cycle moves among what do they adapt and why (climate etc.); How do they adapt (processes); What impacts result? (ecosystem, socioeconomics, integrated systems)

- Understanding impacts and risks: Recognizing the specific climate hazards (increased heat, droughts, floods) and the risks they pose to a particular area (geographical space) is the crucial first step. This involves understanding not only the changes already happening but also anticipating future impacts across varying time horizons (short-term reactive, long-term anticipatory)
- Developing and selecting adaptive options: This involves considering a range of potential solutions (social, technical, behavioral, infrastructural, etc.) to address identified risks, considering available resources, local context, and political feasibility within a given area
- Implementation: Putting the selected adaptation strategies into practice, requiring the activation and allocation of resources. The scale of implementation varies, from local community-based actions to national-level policies and infrastructure projects
- Monitoring and reassessment: Continuously tracking the effectiveness of adaptations, learning from successes and failures, and adjusting strategies as climate impacts evolve and knowledge improves over time. This iterative process acknowledges the dynamic nature of climate change and the need for flexible and responsive adaptation pathways
- Essentially, adaptation is a dynamic process of continuous learning and adjusting to the changing climate, with tailored strategies applied across different geographic locations and time scales

9.10 Adaption Framework for Changing Climate

- Climate change adaptation framework provides structured approaches to reduce vulnerability and enhance resilience to the impacts of a changing climate. The framework guide decision-making and actions at various levels – from individuals and local communities to national governments and international organizations **(Fig:9.3)**

- The framework takes into consideration
  - Adaptation: Stakeholders; traditional, current and future knowledge-information-data-models
  - Adaption science: Research/scenarios; monitoring/data management; assessment
  - Adaptive management: Socio-economic sectors; ecosystem/species; integrated assessments
  - Adaptation options: Reduce vulnerabilities; enhance opportunities; options; assessments
  - Both adaption science and adaption options come under "Science" (understanding prediction)
  - Adaptive management and again adaption options come under "Society" (communication and education

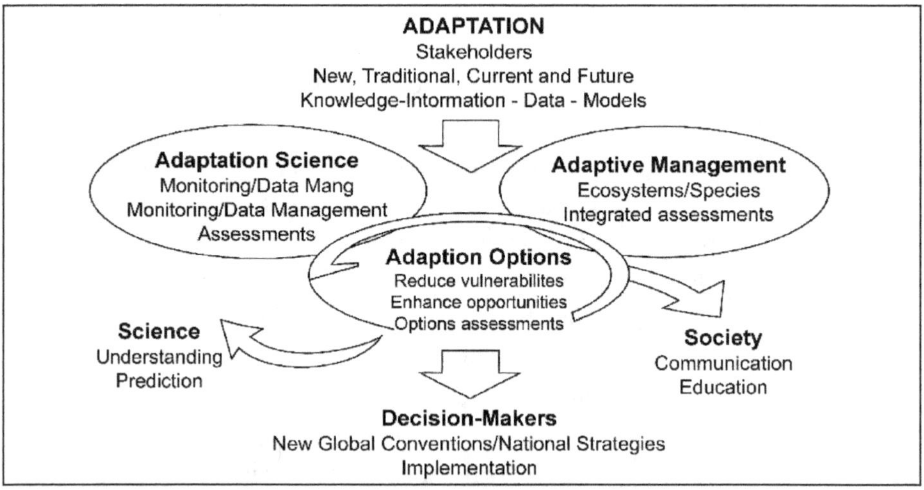

Fig. 9.3 Adaption framework for changing climate.

Source: Unknown. Class lecture notes collection

- Core principles and objectives
  - Reduce risk and vulnerability: Identify potential climate change impacts and prioritize actions to minimize harm to communities, ecosystems, and infrastructure
  - Strengthen resilience: Build the capacity of systems to cope with, recover from, and adapt to climate change impacts

- Integrate with sustainable development: Align adaptation actions with broader development goals and policies for long-term effectiveness
- Emphasize equity and inclusion: Ensure that adaptation measures benefit all members of society, particularly vulnerable groups, and do not exacerbate existing inequalities
- Promote collaboration: Foster cooperation among various stakeholders, including governments, communities, businesses, and scientists, for effective and sustainable adaptation

- Key components
  - Risk assessment and vulnerability analysis: Identifying climate-related hazards, assessing exposure and sensitivity of populations and assets, and evaluating potential consequences
  - Adaptation planning and decision-making: Developing strategies, policies, and measures to reduce climate risks and enhance resilience, including selecting the most suitable options and outlining implementation plans
  - Implementation strategies: Putting adaptation actions into practice, which can include a range of measures like infrastructural and technological changes, institutional adjustments, and behavioral shifts
  - Monitoring and evaluation: Tracking progress, assessing effectiveness, and identifying areas for improvement to ensure that adaptation efforts remain relevant and successful
- Challenges
  - Uncertainty and complexity: Dealing with the inherent uncertainties of climate change projections and impacts
  - Financial and resource limitations: Securing adequate funding and resources for adaptation, especially for vulnerable nations and communities
  - Institutional and governance barriers: Overcoming challenges in coordination, policy integration across sectors and levels of government, and fostering strong political will

- Social and cultural considerations: Addressing diverse needs and priorities within communities, ensuring equitable outcomes, and potentially overcoming resistance to change
- Effective adaptation frameworks are crucial for building resilient communities and ecosystems in the face of climate change

9.11 The Concept of "Climate Neutral"

- Definition: "A way which produces no net greenhouse gas (GHG) emissions". This is achieved by reducing greenhouse gas emissions as much as possible and using carbon offsets to neutralize the remaining emissions
- The issue of climate change stays until it is appropriately addressed. "Climate neutral" future is one of the major options. It takes patience, persistence, and determination. However, it can be done
- The journey to climate neutrality is not a straight line, but a cycle, a matter of cutting down greenhouse gas emissions responsibly and offsetting the remainder

9.11.1 Reasons for Climate Neutral

- There are four reasons for climate neutral and reducing climate footprint
  - Sparing the climate
    - The buildup of GHGs threatens to set the earth on the path to a different climate
    - The IPCC observed that the globe is warming, and it will be warmer than at present
    - Droughts, floods, storms, etc., will become more frequent threatening food supplies
    - Plants and animals which cannot adjust may become extinct

    - Sea levels are rising and will continue to do so, forcing hundreds of thousands of people in coastal zones to migrate
    - Scientists emphasize that the earth's average temperature should not rise by more than 1.5 degrees centigrade over pre-industrial levels
- Conserving natural resources
    - There is growing evidence of another and quite different threat developing. Humans may soon run short of the fossil fuels (gas and oil) which keep modern society going due to natural deflection
    - By 2030, world energy use will probably have increased by more than 50 per cent. Earth can attain energy security when it moves from fossil fuels to fossil free alternatives
- Protecting human health
    - Emissions linked to burning fossil fuels (SO_2, NO_2, etc.) often make people ill
    - Three million people die every year due to outdoor air pollution
    - Pollution comes from vehicles, power stations, and factories
    - It also damages natural world through acid rain and smog
- Boosting economy
    - Individuals who reduce their energy consumption and its climate impact also save money
    - On a more macroeconomic level, economic opportunities arise from measures taken to reduce GHGs emissions
    - Insulating buildings will not only save energy costs but also give the building sector an enormous boost and create employment
    - While some sectors might suffer increased costs, many will seize the opportunity to innovate and get a step ahead of their competitors in adapting to changed market conditions

9.12 The Concept of "Net Zero Emissions"

- Definition: The balance between the amount of greenhouse gases (GHGs) released into the atmosphere by human activities and the amount removed
- Net zero emissions is a vital concept in the global effort to address climate change
- To achieve net zero, GHG emissions must be reduced to as close to zero as possible, and any remaining emissions must be removed from the atmosphere
- This requires offsetting any remaining emissions through activities like reforestation or carbon capture and storage technologies
- Importance of Net Zero Emission
    - Halting global warming: Global warming is directly related to the cumulative amount of carbon dioxide (CO_2) emissions released into the atmosphere. Sustaining net zero CO_2 emissions is essential for stabilizing warming at any level
    - Limiting temperature increase: The IPCC and the Paris Agreement emphasize limiting global temperature rise to 1.5°C above pre-industrial levels to avoid the most severe climate change impacts. Achieving net zero emissions, particularly for CO_2, around mid-century is crucial for this goal
    - Mitigating climate impacts: Successfully achieving net zero can reduce the frequency and intensity of extreme weather events like heatwaves, heavy rainfall, and droughts. It can also help stabilize sea levels, protect terrestrial and marine ecosystems, and improve human health by reducing air pollution
- Strategies to Achieve Net Zero
    - Reaching net zero requires a comprehensive and multi-pronged approach
    - Deep and rapid emission reductions: This involves transitioning from fossil fuels to clean energy sources like solar and wind power, improving energy efficiency across all sectors (transportation, industries, and buildings)

adopting sustainable agricultural practices, and managing waste effectively. It is essential to strengthen the Nationally Determined Contributions (NDCs)

- Carbon removal: Since eliminating all emissions completely is difficult, removing CO_2 from the atmosphere becomes vital to achieve net zero. This can be achieved through
- *Nature-based solutions*: These include reforestation, afforestation, soil carbon enhancement, and sustainable land management practices
- *Technological solutions*: This includes technologies like “Direct Air Capture and Storage” (DACS), which directly capture CO_2 from the atmosphere and store it permanently. Bioenergy with Carbon Capture and Storage (BECCS), which combines biomass energy production with carbon capture
- Aligning short-term actions with long-term goals: Countries need to ensure that near-term emissions reduction targets (like those in NDCs) align with their longer-term net zero commitments, typically set for around 2050. This prevents locking in carbon-intensive infrastructure and technologies that would hinder future progress
- International cooperation and equity: Achieving net zero globally requires strong international collaboration, with developed countries taking the lead

- Challenges
  - While the momentum for net zero is growing, challenges exist
  - Pace of progress: Even though there is excellent progress, the world is still not fully on track in the desired ways to meet the 1.5°C target
  - Feasibility and cost: The scale of transformation required is immense, raising concerns about the feasibility and cost of implementing these changes across all sectors, particularly for developing nations

- Looking ahead
    - The transition to a net zero economy is a massive undertaking, but the benefits of a stable climate, cleaner environment, and new economic opportunities make it a critical and necessary endeavor
- Moving forward. It is crucial to,
    - Ensure the robustness of net zero commitments: Emissions reduction targets need to be ambitious, near-term, and avoid over-reliance on offsets
    - Invest in and scale up clean technologies and carbon removal solutions: Many of the necessary technologies are already available, but require further investment and development
    - Promote international cooperation and equitable solutions: Sharing best practices and technologies and ensuring that the transition is fair and inclusive are crucial for global success
    - Ultimately, achieving net zero is about creating a sustainable future and requires concerted effort from governments, industries, and individuals worldwide

9.13 Climate/Carbon Neutrality Vs Net-Zero

- Worldwide academicians recognize “Climate neutrality” is synonymous with “Carbon neutrality”
- While often used interchangeably, there are subtle differences between "Net-zero emissions" and "Climate/ carbon neutrality".
- Some definitions of carbon neutrality might allow for a greater reliance on carbon offsetting, whereas net-zero emphasizes deeper emission reductions and a more comprehensive approach to carbon removal

9.14 Urgency of Near-term Integrated Climate Action

- Climate change is a threat to human wellbeing and planetary health. Climate resilient development integrates "Mitigation and Adaptation" to advance sustainable development for all and are enabled by increased international cooperation
- This cooperation is essential in improving access to adequate financial resources particularly for vulnerable regions, sectors, groups, inclusive governance, and coordinated policies
- The choices and actions implemented in this decade will have impacts now and for thousands of years
- Harnessing renewable energy from solar, wind, biomass, hydro power, and geothermal plants can reduce carbon/ climate footprints as well

9.15 Addressing General Climate Change Issues

- The individuals, families, communities, governments, organizations, etc., have the best sciences available to plan for and implement climate-smart and resilient practices. Two critically important approaches which are available to managing climate resilience are mitigation and adaptation. These strategies are inter-related and must be pursued simultaneously as both can potentially improve resilience to the changing climate.

Chapter 10

Climate Change and Global and Indian Agriculture

"Agriculture, deeply rooted in Vedic wisdom, reveres Earth as the Mother, emphasizing mindful cultivation and sustainable practices. Through harmonious coexistence with nature and the nurturing of all life, Vedic farming aims to replenish the Earth's vitality and maintain ecological balance for future generations. Honoring the sacred interdependence of all beings, it ensures a healthy atmosphere and a flourishing Earth, embodying the principle of Dharma in action"

10.1 Introduction to Climate Change and Global and Indian Agriculture

- In addition to the points mentioned in Chapter 1 of the book,
    - Climate change has a direct influence on the quantity and quality of agricultural production
    - This is more pronounced in developing countries, where technology generation, innovation, and adoption are slow to counteract the adverse effects of climate change
    - The interdisciplinary nature of these issues requires a long-lasting and where possible more substantial role for

agrometeorology in the efforts to promote sustainable agricultural development during the 21st century

- There is a need to develop locally new and better agro-meteorological adaptation strategies to increasing climate change, especially in vulnerable regions where agriculture is most sensitive to climatic change
- Climate affects many aspects of plant growth and yield and hence the effects of climatic elements and their extremes will significantly alter the crop productivity and in turn, the socioeconomic conditions of many societies, both developing and developed
- Ambient carbon dioxide, light and moisture supply affect crop growth while the rate of crop development is determined by temperature and day length
- Water supply affects both growth and development
- The final crop yield is the integrated result of both growth and development
- Agriculture in low latitudes may be especially affected, where crop yields are projected to decrease even for small amounts of warming
- For mid and higher latitudes, even without adaptation, the effects of warming on wheat yields are projected to be beneficial for local temperature increases
- Adaptation aims at reducing the risk and damage from current and future harmful impacts cost- effectively or exploiting potential benefits
- Agriculture is the world's single largest employer and most climate and weather dependent of all human activities
- The agriculture sector is crucial for the national economy of many countries and the livelihoods of a large population

- It contributes to food and nutritional security, health, and agri exports
- Agriculture is a significant sink of carbon dioxide and at the same time it is also a source of greenhouse gases
- Agriculture is a source of methane and nitrous oxide
- Therefore, agricultural adaptation and mitigation measures vis-a-vis climate change are inter linked

10.2 IPCC Report on Impact of Climate Change on Agriculture

- Changes in Crop Production and Adaptation Strategies
    - Climate change significantly impacts agriculture, creating challenges for food security, farmers, and the broader environment
    - These impacts are multifaced and interlinked viz., reduced soil fertility, alterations in nutrients dynamics, alteration in plant pest relation, low crop productivity, increase in evapotranspiration leading to water loss, alteration in soil moisture relations **(Fig:10.1)**
- Changes in Crop Production

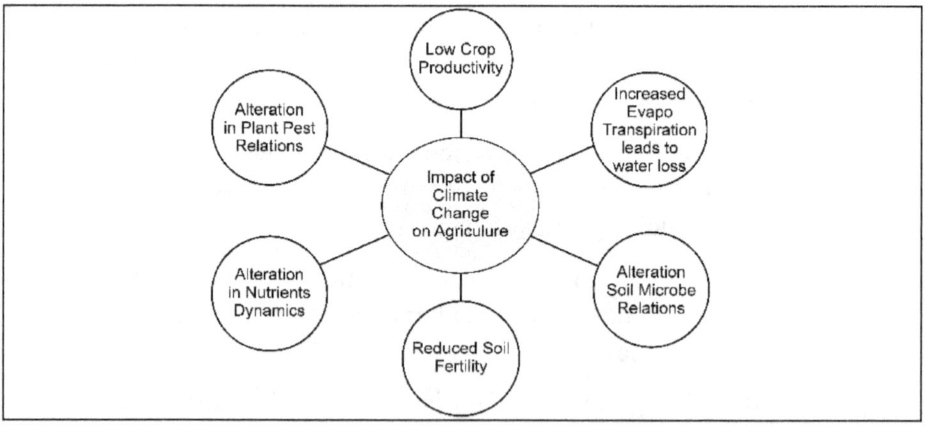

Fig. 10.1 Schematic representation of impact of climate change on agriculture.
Source: IPCC 2021, 2022, and 2023

- Reduced soil fertility and alterations in nutrients dynamics: Changes in precipitation can lead to soil erosion and loss of essential nutrients thereby soil fertility. This in turn may lead to unfavorable dynamics of soil nutrients. Elevated carbon dioxide levels can reduce protein and mineral content in crops potentially affecting food quality and nutrition
- Plant pest relation: Rising temperatures, altered precipitation patterns (leading to droughts or floods) and an increase in pests and diseases are projected to decrease yields for major crops, particularly in warmer regions
- Low crop productivity: Climate change creates less predictable growing seasons, increases the risk of extreme weather events, and could lead to greater instability and low crop production
- Increase in evapotranspiration leading to water loss, alteration in soil moisture relations: Droughts can create or exacerbate evapotranspiration resulting in water shortages for irrigation and overall agricultural use. Alterations in soil moisture relations may also occur

- Adaptation Strategies
  - Farmers are adapting by adopting drought-tolerant crops, diversifying crops and livestock, and utilizing precision agriculture technologies
  - Governments and organizations are supporting climate-smart agriculture, researching resilient crops, and providing training and resources to farmers
- Climate change poses significant threats to agriculture by disrupting established growing conditions, increasing production risks, impacting food quality, and affecting the livelihoods of those dependent on farming. Addressing these challenges requires a multifaceted approach involving emissions reduction, mitigation and adaptation strategies and support for vulnerable communities

10.3 Impacts of Climate Change on Global Agriculture

10.3.1 Tropical Regions

- Roughly 40 per cent of the world population lives in the tropics and agriculture is a very important sector for security and the economies of most countries in the tropics
- The current scenario of enhanced temperatures and increased frequency of extreme events, due to climate change, effect the agriculture significantly in the tropics
- The arid and semi- arid regions account for approximately 30 per cent of the world total area and are inhabited by approximately 20 per cent of the total population in Africa, 17 per cent in the Americas and the Caribbean, 23 percent in Asia, 6 per cent in Australia and Oceania, and 11per cent in Europe
- The agricultural climate of the arid and semi-arid tropical regions in Asia, Africa and Latin America, is characterized by low and variable rainfall and consistently high temperatures during the growing season
- Impacts of climate change are common in these regions with traditionally low agricultural productivity
- The projected climate change and the attendant impacts on water resources and agriculture in the arid and semi-arid tropical regions add additional risk and uncertainty to agricultural systems that are already affected by land degradation due to growing population pressures
- In certain agro-ecological zones, where the predominant soils are sandy in nature, increased mean temperature could affect the maximum temperatures at the soil surface substantially
- Surface soil temperatures could exceed even 60 degrees and under such temperatures, the plant enzymes degrade and limit photosynthesis and growth

- Increased temperatures also result in increased rates of potential evapotranspiration
- In the long term, the very establishment and survival of species in both the managed and unmanaged ecosystems may be threatened, resulting in a change in the community structure
- The prospect of global climate change has serious implications for water resources and regional development
- Projected temperature increases are likely to lead to increased open water and soil/plant evaporation
- The humid and sub-humid tropical conditions are found in nearly 50 per cent of the tropical land mass and 20 per cent of the earth's total land surface
- Tropical Central and South America contain about 45 per cent of the world's humid and sub-humid tropics; Africa about 30 per cent, and Asia about 25 per cent
- Climate change negatively impacts agricultural production in lower latitude and lower income countries. Also, climate change, particularly in terms of frequency/ intensity of droughts, have larger impacts on the sub-humid than on the humid regions.
- If climate change induced disasters become more common, widespread and persistent, many countries in the humid and sub-humid tropical regions will have difficulty in sustaining viable agricultural practices
- Africa
    - It is estimated that the potential evapotranspiration over Africa is projected to increase by 5-10 per cent by 2050
    - Since Africa is the continent with the lowest conversion factor of rainfall to runoff (averaging 15 per cent), and rainfall in some areas are projected to decrease, the predicted dominant impact of climate change is, a reduction in soil moisture in sub-humid zones. However, a reduction in runoff may also occur which benefits agriculture

- The general conclusion is that climate change will affect negatively in some parts of Africa and positively enhance prospects for crop production in other areas. Expansion of agriculture is important in the east African highlands. For example, agroecological suitability in the highlands of Kenya would increase by 20 per cent with warming of 2.5°C based on an index of potential food production
- In Africa, most mid-elevation ranges, plateaus, and high – mountain slope areas are under considerable pressure from commercial and subsistence farming activities. Mountain environments are potentially vulnerable to the impact of increased temperatures. This vulnerability has important ramifications for a wide variety of human uses

- Asia
    - In tropical Asia, more than half of the labor force is employed in agriculture, accounting for as low as 10 per cent and as high as 63 per cent of the GDP in countries of the region
    - The water and agriculture sectors are likely to be most sensitive and hence vulnerable to climate change induced impacts in arid and semi-arid tropical Asia
    - Croplands in many of the countries in the region are irrigated because rainfall is low and highly variable
    - The agriculture sector in Asia is highly vulnerable to climate change because of degradation of the arable land
    - The predicted increase in frequency and/or severity of extreme events coupled with any increase in intensity of tropical cyclones could further exacerbate adverse impacts of climate change on the agricultural sector
    - Studies under elevated CO_2 conditions suggest that areas in mid and high latitudes will experience increases in crop yields, whereas yields in areas in the lower latitudes will decrease

- The climate change endangers sustained agricultural production in tropical Asia in coming decades
- The scheduling of the cropping season as well as the duration of the growing period of the crops would also be affected
- Studies conducted on rice in India, Indonesia and the Philippines confirmed that spikelet sterility and reduced yields negate any increase in dry-matter production because of CO_2 fertilization
- Rice yields could decline by 1per cent annually because of increases in temperature

- Latin America
  - Climate change is affecting agriculture in Latin America
  - Periodic occurrences of severe EL Nino-associated droughts have resulted in occasional famines
- Mitigation and adaptation in tropical regions
  - Monitoring crop growth and development together with appropriate weather and climate information will improve management
  - Development of water conservation strategies, both from traditional and modern practices, is recommended for more efficient water usage
  - Increase the planting of shelterbelts or the use of scattered trees amongst crops for the reduction of erosion and wind damage and conservation of moisture
  - Implementation of both sustainable agriculture, and forestry practices conserve land and improve yields over long term
  - Development of innovative new technologies (e.g., climate forecasting) alongside traditional methods (e.g., intercropping, mulching) improve crop yield
  - Development of adaptation strategies such as response farming at the local community level engage active participation of the land users

10.3.2 Temperate Regions

- Climate change resulted in greater fluctuations in crop yields during recent decades in temperate regions
- Extreme weather events such as drought, flooding, and heat waves resulted in severe impacts on agriculture and forestry, changes in drought tendencies, soil moisture availability and frost-free growing seasons
- Agriculture has also played a role in greenhouse gas emissions. The clearing of forests, draining of wetlands, and the ploughing of rangelands have led to a significant increase in atmospheric CO_2, as organic carbon was decomposed
- Nitrous oxide (N_2O) originates as a by-product of nitrogen fertilizer application and from water logged soils
- Thus, in higher latitudes, a spring burst of Nitrous oxide emissions occurs with rapid snowmelt
- Heavy rains in low-lying areas also cause a Nitrous oxide burst of emissions
- Methane (CH_4) emitted from agriculture is produced by the microbial breakdown of plant material and in the digestive system of cattle
- Distinct regional patterns by latitude were discernible in future climate scenarios for Europe and North America
- Temperatures are projected to increase in nearly all areas but the largest temperature increases are projected over southern portions of both the United States and Europe
- Consequently, the extreme cold of winter is expected to diminish but a greater likelihood of heat waves is projected in summer
- An increase in the frequency and intensity of heavy precipitation is expected, even in southern Europe and the southern United states, despite projections of total precipitation to decrease

- Northern crop areas of both Europe and the United States are projected to have a longer growing season and an expansion of suitable area for crop production
- With higher crop production, however, the increased risk of nutrient leaching and an accelerated breakdown of soil organic matter may affect the quality of northern agricultural lands
- Lower crop yields are expected in southern crop areas due to the warmer and drier summers
- Mitigation and adaptation in temperate regions
    - Earlier planting and sowing of crops are recommended as increase in temperatures are projected. This will utilize higher temperatures earlier in the growing season and result in conserving soil moisture
    - Earlier planting with the use of long season varieties to increase crop yields is a recommended adoption strategy in cooler climate regions where soil moisture is adequate and the risk of heat stress is low
    - In hotter regions, the introduction of shorter season varieties provides a measure to avoid or reduce summer heat and water stress
    - Introducing changes in land allocation for stabilization of crop production and conservation of soil moisture is essential
    - Plans are essential for shorter crop rotations and crop thinning in areas that experience higher precipitation
    - Reduce the impacts of drought and erosion by utilizing larger spacing in forestry plantation planting and later thinning
    - Increase in application of integrated nutrient and pest management techniques

10.4 General Mitigation and Adaptation options for Global Agriculture

10.4.1 Options for Mitigation

- Prevention of overgrazing of grasslands will produce a moderate improvement in soil carbon levels
- Allocation of fallow summer areas is recommended for pasture and rangeland agriculture so as to reduce emissions of nitrous oxide
- Introduction of reduced tillage intensity and summer fallow areas, improved manure management and feed rations with improved drainage and irrigation will all contribute to less emissions of carbon dioxide, methane, and nitrous oxide
- Introduction of forage cropping into rangeland and pasture rotations can be applied for increasing the carbon sequestration
- Improve nutrient management with suitable fertilizers aids the sequestration of carbon

10.4.2 Options for Adaptation

- Climate change and inherent climatic variability make adaptation unavoidable. These are embedded in issues such as sustainability of land productivity, changes in erosion, degradation and environmental quality, which also require due consideration
- Improved management strategies are required for coping with increasing climate change and variability. These can be drawn from both traditional and new technologies
- standardization of crop models for widespread agrometeorological applications are needed, with more modeling on the rainfall distribution and commencement for the rainy season in tropical and subtropical regions
- Changes in agronomic practices, such as earlier planting dates or cultivar substitution, and methods of microclimatic modification, for example to cool animal environments as the climate warms

- Development of physiologically based animal models with well-developed climate components are needed urgently to cover gaps in knowledge and for future projections
- Improvement of carbon sequestration is required by agriculture and forestry by adopting permanent land cover, utilizing conservation tillage, reducing fallow land in summer, incorporating rotations of forage and improving nutrient management with fertilizers

10.5 Nature-based options for Mitigation and Adaptation for Global Agriculture

- Nature-based solutions (NBS)/ actions are those that protect, sustainably manage and restore natural or modified ecosystems which address societal challenges effectively and adaptively, simultaneously providing human well-being and biodiversity benefits
- Optimizing grazing: Optimum grazing efficiency exploits the inherent capacity of grasses (and legumes) to recover following defoliation.
- Improved rice cultivation: Water management of rice fields can reduce methane emissions by as much as 90 percent compared to full flooding. The practice improves rice yields.
- Crop nutrient management: This involves the proper rate of application, applying the proper type of nutrients appropriately and at the correct time.
- Biochar: Biochar is a charcoal-like substance that is made by burning organic material from agricultural wastes by pyrolysis.
- Conservation agriculture: This sustainable farming method is based on crop diversification, minimal soil movement and permanent soil cover.
- Trees in crop plants: Trees store vast amounts of carbon, and provide shelter prevent erosion, diversify production and maintain moisture levels.

- Restoring wetlands and peat lands: Peat wetlands are super carbon sinks. They hold twice as much carbon as all of the world's forests combined, and cover about 3 per cent of land surface of the earth.
- Fire risk management: A bushfire management plan shows all of the bushfire protection measures that will be implemented to reduce the risk for agriculture
- Climate change is on the increase with both global (El Nino/Southern Oscillation) and regional (North Atlantic Oscillation, Southern Annular Mode) modes
- Temperatures have increased with a significant increase in heatwaves, and increases in extreme rainfall, with regional increases in drought and dry spells and intensification of dry seasons and drought
- Increases in pest and diseases, aridity and heat stress have decreased yields of many crops, especially in subtropical, monsoonal and Mediterranean regions owing to global warming
- Adaptation measures include improved seeds, soil and water management, change in sowing dates of crops and changing crops and livestock with diversification
- Largest improvement in resilience occurs with these nature-based solutions

10.6 Potential Agricultural Mitigation and Adaptations to Climate Change at various Scales

- According to the Guide to Agricultural Meteorological Practices (GAMP) of The World Meteorological Organization, Geneva, Switzerland examples of potential agricultural mitigation and adaptations to climate change at various scales are as follows **(Table:10.1)**

Table 10.1 Examples of potential agricultural mitigation and adaptations to climate change at various scales (Source: WMO, GAMP)

Scale				
Global	**National**	**Regional**	**Local**	**Farm**
Shifting centers of production Plant and animal breeding for heat and drought tolerance	-Land allocation -Labor supply and demand -Balance of food and non-food crops -Policy to support farm level adaptation -Plant and animal breeding for heat and drought tolerance	-Land allocation -Labor supply and demand -Balance of food and non-food crops -Policy to support farm level adaptation -Plant and animal breeding for heat and drought tolerance	-Type of farming - Rotations - Crop "Mixes" - Balances of cash vs food crops - Water management + variety selection, animal breed selection, timing of activity (DOS), water conservation	- variety selection -Animal breed selection - Timing of activity. (DOS) -Water conservation

10.7 Climate Change and Indian Agriculture

- The GHG emissions, surface average temperature, and sea level are projected to rise due to climate change on the Indian sub-continent
- It promotes increasingly stronger impacts on Indian agriculture and forestry
- Presently agronomic adaptation has been effective, however, there is a need for interdisciplinary approaches
- The negative impacts of climate change necessitate strategies, and interventions thorough a wide range of mitigation and adaptation measures and options for continued success of Indian agriculture

10.8 GHGs Emission Scenarios of India

- The current GHG emission scenario of the world and India indicate that in 2020, the world emitted about 50 billion tons carbon dioxide equivalent GHGs globally. In this, world agriculture alone contributes 10-12 per cent. Interestingly, agriculture, forestry and other land use (AFOLU) together contributes 24-25 per cent of total emissions
- India emits 6 per cent of global GHGs. In this the Indian agriculture contributed 0.8 per cent to global emissions. Currently, total GHG emissions is 2.82 billion tons carbon dioxide equivalent per annum from all sectors of Indian economy. The energy sector contributes 75 per cent of these emissions followed by agriculture 14 per cent, industry 8 per cent and waste 3 per cent
- In quantitative terms, India's agriculture emits 471 million tons carbon dioxide equivalent. Of these emissions 55 per cent comes from enteric fermentation, 19 per cent from soil (mostly nitrous oxide), 17 per cent from rice fields, 7 per cent from manure management, and 2 per cent from residue burning and other sources
- In the crop sector, which emits about 40 per cent of the total, low land rice emits the largest amounts of GHGs, followed by sugarcane and potato

10.9 Mitigation Targets and Commitments by the Government of India

- At COP 26 of UNFCC in Glasgow, UK, in November 2021 India committed to achieve the target of net zero emissions by 2070 and has made five commitments "Panchamrit commitments". These include that by 2030, India will reduce emissions intensity by 45 per cent. It is proposed to draw 50 per cent of energy from renewable resources. Also, India proposed to raise non fossil fuel-based energy capacity to 500 giga watt and cut carbon emissions by one (1) billion tons by 2070 such that the country will attain net zero emission and carbon neutrality

- In addition, India submitted its long-term low carbon development strategy document in November 2022 which provides seven "Transitions" for achieving net zero emissions. These transitions, however, do not include the Indian agriculture sector. Indian agriculture was purposefully excluded because it is linked to food security and the livelihoods of millions in India. In addition, agriculture is practiced by small and marginal farmers, many of them having small land holdings unlike the industrialized western countries
- Also, for the voluntary emission intensity reduction goal announced by India at Copenhagen (2009) and subsequently in the NDCs (2015 & 2022), the emissions from the agriculture sector have been excluded. India has rightly taken the stand that mitigation burden should not be transferred to its farmers
- Even though the Indian agriculture was excluded in India's low carbon strategy, to fulfil the commitment of Net Zero emissions by 2070, and to feed its burgeoning population under the impending negative impacts of climate change scenarios appropriate mitigation and adaptation strategies have to be developed and followed by Indian agriculture sector

10.10 Opportunities for Net Zero Emissions by 2070 by India

- Carbon Emission and offsetting strategies in Indian Agriculture Sector are
    - Carbon emission reduction avenues: Solar energy-based ground water lifting, Fertilizer management and cover crops, efficient livestock system, shifting to efficient rice systems of cultivation, food loss minimization, solar / green energy based farm mechanization etc.
    - Carbon offsetting avenues: Agroforestry, rehabilitation of degraded lands, perennial and horticultural crops, circular carbon economy models, waste recycling and biochar, shifting to green energy value chain

10.11 Food needs and associated Technology efficiency Targets by Amritcal (2047)

- Changing Agri-Food System: Now and 2047: Increase in Nitrogen use efficiency, water use efficiency, reducing post-harvest losses, farm mechanization, developing nutri coarse cereals, production increase in food grains, egg, meat, fish, milk, vegetables, fruits, etc.

10.12 Key components of India's Net Zero Commitment

- Emission reduction
- Carbon removal
- Climate resilience
- Fulfilling the Paris Agreement goals
- Ambition and urgency
- Action and accountability

10.13 Key Role players in India's Net Zero Carbon Commitment

- Education, training, and sensitization
- Public policy support
- Voluntary / incentives for environmental services
- Regulatory
- Carbon credits
- Corporates and industry
- NGOs
- Village institutions
- Common interest crops

10.14 Indian Agriculture Sector in achieving Net Zero GHG emission Target by 2070

- Though the Paris Agreement, target commitments consisting of the energy sector along with others are moving in a desired way, the agriculture sector also has to contribute to net zero carbon commitment by 2070
- Considering several climate actions and national programs, as well as State climate adaptation programs in agriculture and allied sectors over the past decade, 471 million tons of carbon dioxide equivalent can be taken as current estimate of GHG emissions from the agriculture sector
- The implementation of smart TIFER is the key to net zero emissions. (a) tillage (conservation tillage); (b) irrigation (micro, sensor based); fertilizer application (demand driven, sensor based); energy (solar); and residue management (recycling and reuse)
- In addition, there is a need for macro scale initiatives such as agroecological planning for agriculture
- Shifting one third rice area by 2047 to DSR and water – nutrient efficiency, rice systems can reduce the larger share of methane and nitrous oxide emissions from the agriculture sector
- Biochar is a big opportunity

10.15 Priority Technologies for Implementation towards Net Zero Commitment in India

- Total agriculture emissions target reduction is 471 Mt Co2- eq
- Livestock emissions 55 per cent current emissions (Green fodder, feed concentrate, shelter and hygiene, tolerant and high yielding breeds etc.)

- Crop production emissions 45 per cent current emissions (paddy, fertilizer, crop residue burning, manure, intensive tillage, water lifting energy, land degradation etc.)

10.16 Available Strategies for GHGs Mitigation in Indian Agriculture Sector

- Indian agriculture has the potential to attain net zero emissions or carbon neutrality, as detailed below, particularly in crop sector in which the emissions are 150 million tons carbon dioxide equivalent per year
- Mitigation through efficient agri-input management
- Mitigation through agri-horticulture systems
- Livestock management
- Agroforestry
- Food value chain: Production, storage, and household consumption
- Agri- Energy, warehouses and local foods in India
- Green foods, vegetarian lifestyle and agri start up innovations
- Waste to wealth (Circular carbon economy, Bio energy and Biochar)
- Replication of carbon positive villages

10.17 Policy Recommendations

- Inclusion of climate education at all levels is a critical step towards fulfilling India's net zero objectives on GHG emissions
- In climate research, it is essential to fund R&D for renewable energy sources including solar, wind, and biofuels

10.18 Crisp Description of Important Agricultural Systems, Techniques, Technologies, and Approaches for Climate Change Mitigation and Adaptation at all Levels

10.18.1 Organic Farming

- Organic farming is an agricultural system that prioritizes ecological balance, biodiversity, and resource conservation while avoiding synthetic inputs like pesticides and fertilizers. It aims to produce food and fiber through natural processes and sustainable practices, promoting long-term weather health, soil health, animal welfare, and environmental protection
- Core Principles
    - Ecological Balance: Organic farming emphasizes maintaining a healthy ecosystem by promoting natural processes and biodiversity
    - Resource Conservation: It focuses on minimizing resource depletion and maximizing the efficient use of resources like water and energy
    - Soil Health: Organic farming practices like crop rotation, cover cropping, and composting are used to build and maintain healthy soil structure, fertility, and microbial activity
    - Animal Welfare: Organic livestock production prioritizes animal health and well-being, with specific standards for housing, diet, and management
    - Biodiversity: Organic farming encourages a variety of plant and animal species, which can enhance ecosystem resilience and pest control
    - Food Safety and Quality: Organic farming aims to produce safe and healthy food through natural methods and reduced exposure to synthetic chemicals

- Key Practices
  - No synthetic fertilizers or pesticides: Organic farmers rely on natural methods to manage pests and provide nutrients to crops, such as crop rotation, cover cropping, and composting
  - No genetic modification: Organic farming prohibits the use of genetically modified organisms (GMOs) in food production
  - Emphasis on natural pest control: Organic farmers use strategies like biological control (introducing natural enemies of pests), trap crops, and habitat management to control pests
  - Restrictions on antibiotics and growth hormones: In organic livestock production, the routine use of antibiotics and growth hormones is prohibited
  - Traceability and certification: Organic products are certified by accredited bodies to ensure they meet specific production and processing standards, guaranteeing transparency and consumer confidence

10.18.2 Natural Farming

- Natural farming, also known as sustainable or regenerative agriculture, is a farming approach that emphasizes working in harmony with nature, minimizing external inputs, and promoting biodiversity. It focuses on leveraging natural processes and resources to cultivate healthy crops and maintain a thriving ecosystem within the farm. Key aspects include minimal tillage, organic inputs, and encouraging natural decomposition
- Core Principles
  - Working with Nature: Natural farming prioritizes understanding and respecting natural processes, rather than attempting to control or manipulate them

- Minimal External Inputs: It minimizes or eliminates the use of synthetic fertilizers, pesticides, and herbicides, relying instead on natural methods to enhance soil fertility and manage pests and diseases
- Soil Health: Natural farming emphasizes building healthy soil through practices like mulching, cover cropping, and encouraging natural decomposition of organic matter
- Biodiversity: It promotes biodiversity through crop diversification, companion planting, and encouraging a variety of plants and animals within the farm ecosystem
- Zero Budget: Some interpretations of natural farming, like Zero Budget Natural Farming (ZBNF), aim to minimize or eliminate the financial burden on farmers by using locally available resources like cow dung and plant extracts for inputs
- Respect for Nature's Role: Natural farming recognizes the farmer's role as a facilitator rather than a controller, allowing natural processes to take their course

- Key Practices
  - No-Till Farming: Avoids tilling the soil, which can disrupt soil structure and microbial life
  - Organic Mulching: Uses organic materials like straw, leaves, or crop residues to cover the soil surface, suppressing weeds, conserving moisture, and enriching the soil
  - Natural Pest and Disease Management: Employs methods like using natural pesticides derived from plants, encouraging beneficial insects, and promoting a healthy ecosystem to manage pests and diseases
  - Intercropping and Crop Rotation: Diversifies crops and rotates them to improve soil health, nutrient cycling, and pest and disease management.
  - Composting: Utilizes on-site composting of organic waste to create nutrient-rich soil amendments

  - Water Management: Focuses on water conservation through techniques like mulching and contour farming, reducing dependency on irrigation

- Benefits
  - Environmental Sustainability: Reduces pollution from synthetic inputs, enhances soil health, and promotes biodiversity
  - Improved Soil Health: Enhances soil fertility, water retention, and carbon sequestration.
  - Reduced Costs: Minimizes the need for expensive synthetic inputs, potentially reducing farmers' financial burden
  - Increased Resilience: Creates more resilient farming systems that are less vulnerable to climate change and other external shocks
  - Healthier Food: Produces food that is free from chemical residues and potentially richer in nutrients

10.18.3 Conservation Agriculture

- According to FAO, conservation agriculture (CA) is a farming system that promotes minimum soil disturbance (by using no-till practices), maintenance of a permanent soil cover and diversification of plant species. It aims to prevent land degradation and regenerate degraded lands by enhancing biodiversity and natural biological processes above and below the ground surface that contribute to increased water and nutrient use efficiency and improved and sustained crop production
- Core Principles
  - Minimal soil disturbance: Eliminates or greatly reduces mechanical tilling, plowing, and seedbed preparation. This practice, often called no-till farming, protects the soil's structure and organic matter, preventing erosion and loss of fertility
  - Permanent soil cover: Maintains a protective mulch layer on the soil surface year-round. This cover can consist of crop residues from previous harvests or specifically planted cover

crops. It shields the soil from harsh weather, suppresses weeds, and maintains soil moisture

    - Crop diversity: Involves rotating different types of crops and using intercropping techniques. This breaks pest and disease cycles that typically thrive in monoculture systems. Planting diverse root structures also improves soil health and nutrient cycling

- Methodology
    - The implementation of CA combines these principles through a variety of specific techniques
    - No-till planting: Specialized equipment, such as no-till seed drills, places seeds directly into the soil through the layer of crop residue with minimal disturbance
    - Cover cropping: Farmers plant non-cash crops (legumes or grasses) in the off-season to provide continuous ground cover, prevent erosion, and add organic matter to the soil
    - Integrated pest management (IPM): By promoting biodiversity and disrupting pest cycles through crop rotation, CA reduces the reliance on synthetic pesticides
    - Precision nutrient management: Nutrients are applied in optimal dosages based on soil tests to minimize environmental impact and not disrupt the biological processes in the soil
    - Controlled vehicle use: Farmers use designated lanes for vehicles to minimize soil compaction across the field
    - Residue management: Crop residues are deliberately left in the field after harvest instead of being removed or burned. This material serves as the permanent soil cover

10.18.4 Carbon Farming

- Definition: A set of agricultural practices designed to remove carbon dioxide from the atmosphere and store it in the soil and plant matter. Also known as "Regenerative agriculture"
- Carbon farming aims to encourage the absorption and long-term storage of carbon in the soil and plants
- The benefits include climate change mitigation, improved soil health etc.
- This involves carbon sequestration "Capturing and storing atmospheric carbon dioxide in soil and vegetation" to offset emissions.
- By integrating carbon farming into national policies and supporting farmers through financial incentives, Nations can build climate resilience while fostering a more sustainable growth
- Carbon farming is emerging as a key strategy in mitigating climate change while promoting sustainable agricultural practices
- Farmers can potentially generate income through carbon credits or by selling carbon-friendly product
- Examples: Cultivation of cover crops, no- till farming, composting, agroforestry etc.,

10.18.5 Climate Smart Agriculture

- Definition: An integrated approach to managing landscapes (cropland, livestock, forests, and fisheries) that addresses the interconnected challenges of food security and climate change
- It aims to sustainably increase agricultural productivity and incomes, adapt to and build resilience to climate change, and reduce and/or remove greenhouse gas emissions
- This is also an adaptive approach aimed to ensure food security by mitigating the adverse effects of climate change on crop production

- Climate Smart Agriculture (CSA) is a transformative approach to agricultural production, acknowledging the realities of climate change
- The CSA manages climate related risks by enhancing the resilience of agriculture and food production systems at various levels
- This involves disseminating advanced weather information to farmers through cutting edge information technologies, such as SMS, mobile apps, email etc., enabling them to minimize crop losses.
- The three pillars of CSA are
  - Enhancing agricultural productivity and incomes
  - Adapting to and building resilience against climate change
  - Reducing or eliminating greenhouse gases from agriculture
- CSA recognizes that these three objectives are interconnected and should be pursued in an integrated manner, rather than in isolation. Examples: Conservation agriculture, Agroforestry, cultivating climate resilient crop varieties, INM, IPM etc.

10.18.6 Climate Resilient Agriculture

- Definition: A set of farming practices designed to mitigate and adapt to the impacts of climate change
- It encompasses a range of strategies aimed at maintaining or enhancing agricultural productivity and food security in the face of a changing climate.
- Adaptation includes adjusting farming methods to cope with climate change effects shifting planting dates, using drought-resistant crops, or improving water management techniques
- Mitigation includes reducing agriculture's contribution to climate change through practices that lower greenhouse gas emissions or enhance carbon sequestration in agricultural lands.
- Sustainable productivity and food security in the face of climate change are the other key aspects

- By implementing these and other CRA practices, farmers can enhance their ability to cope with climate change impacts, ensuring food security and promoting sustainable agricultural systems
- Examples of CRA practices: Water management practices like rain water harvesting, drip irrigation, mulching etc.; Soil management practices like cover cropping, crop rotation, no- till farming; IPM, INM; Agroforestry, Crop diversification

10.18.7 Controlled Environment Agriculture (CEA)

- Definition: Cultivation of crops within enclosed structures where environmental conditions are precisely managed
- This farming allows for year-round crop production, optimized yields, and reduced reliance on pesticides and herbicides
- Key aspects of CEA are controlled environment, enclosed structures, advanced technologies, optimized production, reduced Reliance on external factors etc.
- The CEA methods are hydroponics, vertical farming, greenhouse systems etc.
- To reduce the agricultural imports CEA interventions are essential
- Limitations: High initial cost; gaps in technical knowledge; unreliable energy systems
- Examples: Greenhouses, vertical farms, indoor farms using hydroponics, aeroponics, or aquaponics etc.

10.18.8 Regenerative Agriculture

- Regenerative agriculture is a way of farming that nurtures and restores soil health, and therefore reduces water use, prevents land degradation, and promotes biodiversity. By minimizing land ploughing, practicing rotating crops, and using animal manure and compost, regenerative agriculture ensures that the soil stores more carbon, conserves more moisture, and is healthier due to thriving fungal communities.

- Intensive agriculture is responsible for a third of global greenhouse gas emissions, uses 70 percent of the fresh water we consume, and leads to soil degradation through its use of heavy machinery, chemical fertilizers, and pesticides. It is also the biggest contributor to biodiversity loss. By contrast, regenerative agriculture helps lower greenhouse gas emissions, conserves water, and restores land. Moreover, healthy soil produces more food and better nutrition and has other positive impacts on ecosystems and biodiversity.

10.18.9 Biochar

- Biochar is a charcoal-like product created by heating biomass (like wood, manure, or agricultural waste) in a limited or oxygen-free environment, a process known as pyrolysis. This heating process prevents the biomass from fully combusting and decomposing into greenhouse gases like methane and carbon dioxide. Instead, it transforms the biomass into a stable form of carbon that can persist in soil for hundreds or even thousands of years
- Key concepts
  - Pyrolysis: The controlled heating of biomass in a low-oxygen environment to produce biochar, bio-oil, and syngas. The specific temperature and duration of this process influence the final characteristics of the biochar produced
  - Carbon Sequestration: Biochar's ability to store carbon in a stable form within the soil, thereby reducing atmospheric carbon dioxide levels and mitigating climate change
  - Soil Amendment: Biochar's application to soil to enhance its physical, chemical, and biological properties, including water retention, nutrient availability, and microbial activity
  - Terra Preta: The concept of using biochar as a soil amendment originated from the observation of remarkably fertile "dark earth" soils in the Amazon Basin, known as terra preta. These soils are believed to have been created intentionally by ancient Amazonian inhabitants through slash-and-char techniques, which involved burying charred plant material

- Benefits of biochar to the soil
  - Improved Water Retention: Biochar's porous structure helps retain water in the soil, making it available to plants for longer periods and potentially reducing irritgation needs
  - Enhanced Nutrient Availability: Biochar can hold onto essential plant nutrients, preventing leaching and making them more accessible to crops, which can potentially decrease the need for synthetic fertilizers
  - Reduced Soil Acidity: Biochar can help increase soil pH, making it more suitable for a wider range of plants and beneficial microorganisms, especially in acidic soils
  - Improved Microbial Activity: Biochar provides a habitat and surface area for beneficial soil microorganisms, supporting nutrient cycling and disease suppression, leading to healthier soil and plant growth
- Environmental significance
  - Climate Change Mitigation: Biochar's ability to sequester carbon in the soil for extended periods is a key benefit, helping to remove CO_2 from the atmosphere and combat climate change
  - Waste Valorization: Biochar production offers a way to convert various organic waste materials into a valuable product, reducing landfill waste and associated greenhouse gas emissions.
  - Reduced Greenhouse Gas Emissions: Applying biochar to agricultural fields has been shown to reduce emissions of potent greenhouse gases like nitrous oxide (N_2O) and methane (CH_4)
- Limitations and considerations
  - The effectiveness of biochar can vary depending on factors such as feedstock type, pyrolysis conditions, soil type, and climate conditions

  - While biochar itself does not add nutrients, biochars made from manure can contribute to nutrient content in the soil
  - The cost of producing and transporting biochar on a large scale remains a challenge, limiting widespread adoption
  - Potential risks associated with biochar use include the possibility of toxicants in certain biochars and the potential for negative impacts on nutrient availability or microbial activity if not used appropriately
- In conclusion, biochar is a versatile carbon-rich material produced through pyrolysis of biomass, offering significant potential for improving soil health, promoting sustainable agriculture, and mitigating climate change

10.18.10 Carbon Offsetting and Carbon Credits

- Carbon offsetting
  - There is only one atmosphere
  - Greenhouse gases circulate freely in the atmosphere
  - Therefore, greenhouse gases emissions saved by one person or in one country are just as valuable as savings by someone else
  - By this way the atmosphere still benefits. So, if someone wants to emit more than they want to or allowed to, they pay to help reduce emissions elsewhere
  - This is known as "Carbon offsetting/trading
  - Example: The climate damage done by a flight journey by an individual may be counteracted by tree plantation to soak up the carbon generated by the journey by the individual
  - In essence, carbon trading is a process of buying and selling carbon credits
  - It emerged as an important strategy in mitigating climate change by incentivizing industries to reduce their greenhouse emissions
  - Carbon, a key element in coal, petroleum, and natural gas contributes to rising GHG emissions, driving climate change

- o The concept of carbon trading was introduced at the 3 rd. COP in Kyoto with the aim to establish an international agreement between member countries to address global warming and GHG emissions
- o It required the Parties to reduce GHG emissions
- o The Protocol made "Carbon" a tradable commodity
- o Carbon trading is done on a virtual platform "Carbon Market"
- o In carbon market two types of traders exist A) Cap and trade which is allowance-based B) Offset trading which is project based and traded under voluntary market
- o The Kyoto Protocol introduced three mechanisms viz.,

  A) International Emissions Trading (IET)

  B) Joint Implementation (JI), and C) Clean Development Mechanism (CDM)

- Carbon credits
  - o When an offset is sold the serial number and "Credit" for reduction is transferred from the account of the seller to an account of the buyer. This is known as "Carbon credit"
  - o The quantification of one ton of carbon dioxide equivalent removed from the atmosphere is represented by carbon credits
  - o These credits are produced by initiatives that either take carbon out of the atmosphere (afforestation projects) or reduce emissions (renewable energy projects)
  - o A carbon credit is gained for each ton of carbon dioxide or its equivalent that is kept out of the atmosphere
  - o The credits are produced through a variety of initiatives, such as methane capture, renewable energy, reforestation, increased industrial efficiency etc.
  - o Carbon credits are significant because they serve as a mechanism based on the market to encourage reductions in emissions

- Kinds of carbon markets
  - Voluntary markets: The trading of carbon credits that occurs voluntarily on the voluntary market is called Verified Emission Reductions (VER). Through the Voluntary Carbon Market (VCM) business and people can choose freely offset their emissions without being governed by International Agreements.
  - Markets for regulatory compliance: Followed by companies and governments which h are required by law to account for their GHG emissions. Carbon credits from these projects are known as Certified Emission Reductions (CER)
  - The Kyoto Protocol established the Clean Development Mechanism (CDM) the object of which was to enable industrialized nations to obtain CER for their expenditure

10.19 Addressing General Climate Change Challenges on Global and Indian Agriculture

- The individuals, families, communities, governments, organizations etc., have the best sciences available to plan for and implement climate-smart and resilient practices. Two critically important approaches which are available to managing climate resilience are mitigation and adaptation. These strategies are inter-related and must be pursued simultaneously as both can potentially improve resilience to the changing climate.

Chapter 11

Climate Change and Major Cereal and Pulse Crops

"Cereals and pulses, as described in the Vedas, are viewed as sacred sustenance, bestowed by Mother Earth. Their cultivation and consumption were part of a holistic philosophy emphasizing responsible use of natural resources and honoring the interconnectedness of life. Through sustainable practices, cereals and pulses contribute to ecological balance and ensure a thriving future for all, embodying the Vedic ethos of environmental stewardship"

11.1 Introduction to Major Cereal and Pulse Crops

- Growing cereals and pulse crops offer numerous benefits for human health, food security, environmental sustainability, and economic prosperity
- Cereals primarily provide carbohydrates for energy, along with fiber, vitamins (especially B vitamins), and minerals
- Whole grains offer significant amounts of fiber and various micronutrients that can help reduce the risk of chronic ill health like heart diseases, cancer, and diabetes
- Cereals like rice, wheat, and maize are staple foods for billions of people worldwide, providing a major source of dietary energy
- Pulses are excellent sources of plant-based protein, dietary fiber, essential vitamins and minerals (Iron, Potassium, and Zinc). They can contribute to healthy weight management, improve

blood sugar control, reduce cholesterol, and potentially lower the risk of chronic diseases including cancer and cardiovascular disease

- When combined with cereals, pulses offer a more complete amino acid profile, fulfilling protein requirements, particularly in plant-based diets
- Pulses are often more affordable than other protein sources, making them crucial for food security, particularly in developing countries

11.2 Major Cereal Crops

11.2.1 Rice (*Oryza sativa*)

- Climatic Requirements
- Rice, particularly Asian rice, thrives in warm, wet climates. The International Rice Research Institute (IRRI) located in the Philippines studies and develops rice varieties adapted to various climatic conditions, including drought, floods, and heat stress
  - o Temperature
    - Rice grows in temperatures above 10°C
    - The optimal temperature for rice cultivation is between 20–35°C
    - Temperatures outside of this range, whether too high or too low, negatively impact growth, physiology, and yield
    - High daytime temperatures and cooler nights during the growing season are generally preferred for economically sound production
  - o Rainfall
    - Rice is a water-loving crop and requires about 10 mm of water daily
    - Traditionally, rice is cultivated in flooded fields, but techniques like Alternate Wetting and Drying (AWD) can reduce water usage significantly without compromising yields

        - Maintaining optimal soil moisture levels during crucial stages like tillering, panicle initiation, and grain filling is vital for maximizing productivity
        - Rainfed rice production is more sensitive to soil moisture variability than irrigated rice
    - Solar radiation/Sunlight
        - Rice requires sufficient sunlight for photosynthesis and development
        - Insufficient sunlight hinders growth, delay in development, reduces grain filling, and affect chlorophyll accumulation leading to yield declines
        - Cloudy conditions during the reproductive stages reduce the amount of sunlight reaching the plants, potentially affecting yield
    - Humidity and wind
        - High humidity and low wind speeds increase the incidence of pests and diseases
- Impacts of climate change
    - Rice is cultivated in more than 100 countries and is a primary food source for over half of global population
    - The crop is vulnerable to climate change there by food security
    - Paddy yields decline under increased temperatures due to climate change
    - Cloudiness induced lower sunshine hours due to climate change results in reduced rice yields
    - Climate change induced significant reduction in rainfall (in upland rice cultivation) during anthesis and grain filling reduces rice yields up to 70 per cent primarily due to lower grain number, weight, and biomass
    - Mitigation and adaptation strategies to climate change
    - Rice crop mitigation and adaptation strategies under global climate change include both reducing greenhouse gas emissions (mitigation) and adjusting farming practices to cope with climate impacts (adaptation). These strategies involve various techniques like alternate wetting and drying (AWD) irrigation, developing climate-resilient rice varieties, and optimizing planting dates

- Alternate Wetting and Drying (AWD): This water management technique involves irrigating rice paddies and then allowing them to dry to a certain depth before re-irrigating. AWD reduces methane (CH4) emissions by introducing aerobic conditions in the soil, inhibiting the activity of methanogenic archaea. It also helps conserve water resources
- Short Duration Rice Varieties: Using rice varieties with shorter growth cycles reduce the number of days the crop is flooded, thus lowering methane emissions
- Developing Climate-Resilient Rice Varieties: Breeding or genetically engineering rice varieties that are tolerant to high temperatures, drought, and other climate change-related stresses is crucial for maintaining rice production
- Reduced Tillage and Organic Matter Management: Practices like reduced tillage and recycling farm waste into organic fertilizers also help mitigate greenhouse gas emissions
- Precision Farming and Reduced Chemical Fertilizers: Optimizing fertilizer use through precision farming techniques can minimize methane emissions from fertilizer application
- Methanotrophs: Introducing microorganisms that oxidize methane can further reduce emissions from rice fields
- Adjusting Planting Dates: Shifting planting dates to avoid periods of extreme heat or drought help mitigate yield losses
- Crop Diversification and Integration: Integrating rice with other crops or farming systems improves resilience and reduces reliance on a single crop
- Weather Forecasting and Climate Information: Utilizing climate forecasting and remote sensing data help farmers make informed decisions about planting, irrigation, and other management practices
- Improved Infrastructure and Support: Providing farmers with access to better infrastructure, financial support, and training on climate-smart practices is essential for successful adaptation

- o Financial Incentives and Subsidies: Governments can provide financial support for farmers to adopt mitigation and adaptation technologies, such as AWD or drought-resistant rice varieties
- o Training and Extension Services: Providing training and extension services to farmers on climate-smart agricultural practices is crucial for effective implementation of these strategies

11.2.2 Wheat (<u>Triticum</u> <u>aestivum</u>)

- Climatic requirements
- Wheat, a staple crop globally, thrives best in temperate regions with moderate temperatures and adequate rainfall
  - o Temperature
    - ▪ Wheat generally prefers warm temperatures (20-24°C or 68-75°F) during its vegetative growth
    - ▪ Germination: 10-30°C (50-86°F)
    - ▪ Tillering: 13-24°C (55-75°F)
    - ▪ Heading & Anthesis: 18-24°C (64-75°F)
    - ▪ Grain Filling: 18-30°C (64-86°F)
    - ▪ However, extreme temperatures, both hot and cold, can negatively impact wheat yields
  - o Rainfall
    - ▪ Wheat is typically grown in areas with moderate rainfall, ideally between 30 and 90 cm annually
    - ▪ Dry spells and drought severely reduce wheat yields, making drought-resistant varieties crucial
  - o Other Climate Factors
    - ▪ Sunshine: Wheat needs ample sunshine, especially during grain filling
    - ▪ Day Length: Wheat varieties are often selected for specific flowering times to avoid late frosts or terminal drought
    - ▪ Disease Pressure: Climate change can also alter disease dynamics, with studies showing that wheat blast could expand its reach and reduce yields

- Impacts of Climate Change
    - o Wheat is a thermosensitive crop, necessitates timely sowing to achieve optimal phenological development, and to enhance biomass production
    - o Climate change often necessitates farmers to change the dates of sowing. Sowing date plays an important role in wheat crop phenology and optimum yields. Adjusting sowing dates mitigate climate risks, enhance resource use efficiency, and improve productivity in diverse agro climatic conditions
    - o Wheat crop sown at optimum date of sowing demonstrate the best growth and yield, as favorable environmental conditions align with the heat unit requirements of different phenophases of wheat crop
    - o Temperature increases during the grain filling stage results in substantial reduction in yields
    - o Raising temperatures due to climate change hinder plant growth resulting in reduced wheat productivity
    - o Wheat yield is highly sensitive to temperatures during reproductive phase
    - o Wheat yields reduce under lesser maximum and minimum temperatures than optimum temperatures
    - o Prolonged dry spells caused by climate change impair plant growth and productivity by disrupting anatomical and physiological processes
    - o Wheat seeds germinate prematurely due to wet conditions caused by prolonged rain or high humidity before harvest
    - o Grain quality and yield are significantly impacted due to climate change induced hailstorms at maturity and harvesting stages of wheat
    - o More number of rainy days have positive impacts at growth and development stages and negative impacts at maturity and harvest stages
    - o Climate change driven higher maximum and minimum temperatures, coupled with lower morning and evening R.H. delays (reduce) onset of yellow rust in heat stress years
    - o Persistently lower temperatures, higher R.H. and rainfall leads to severe yellow rust

- Climate change can also alter disease dynamics, with studies showing that wheat blast could expand its reach and reduce yields

- Mitigation and adaptation strategies to climate change
  - Reducing tillage: Minimizing soil disturbance reduces carbon emissions from the soil
  - Improving nitrogen use efficiency: Optimizing fertilizer application can reduce nitrous oxide emissions, a potent greenhouse gas
  - Promoting carbon sequestration: Practices like cover cropping and conservation tillage increase carbon storage in the soil
  - Breeding for lower emissions: Developing wheat varieties with lower nitrogen requirements or improved nitrogen use efficiency can reduce the need for fertilizers
  - Adjusting sowing dates: Advancing/ delaying planting, depending on location, help wheat avoid heat stress during critical growth stages
  - Developing heat-tolerant varieties: Breeding and selecting wheat varieties that withstand higher temperatures is crucial
  - Integrating climate forecasts: Using weather forecasts to make informed decisions about planting, irrigation, and other management practices can reduce risks
  - Utilizing climate-smart agriculture (CSA) techniques: CSA practices, such as conservation agriculture, agroforestry, and integrated nutrient management can enhance both mitigation and adaptation

11.2. 3 Maize (*Zea mays*)

- Climatic Requirements
- Maize, a globally important cereal, is cultivated under diverse climatic conditions, ranging from tropical to temperate regions and altitudes from sea level up to 4000 meters. Its wide adaptability is due to the availability of cultivars suited to different environments and growing periods. Maize is a warm-weather crop that does not tolerate frost at any stage of growth

- Temperature
  - Optimal temperature range for growth: 20 to 30 °C
  - Optimal daytime temperatures: 20 to 33 °C
  - Optimal nighttime temperatures: 17 to 23 °C
  - Lethal temperature: 46 °C
  - Maize growth decreases at temperatures above 30°C and ceases above 40°C
- Sunlight
  - Full sunlight is preferred, ideally at least 8 hours of daily direct sunlight
  - Maize can thrive in partial sun but performs best under full sun conditions
  - The most beneficial period for sun exposure is between 10 AM and 4 PM
- Rainfall
  - Total seasonal rainfall requirements range from 500 mm to 800 mm
  - Maize is sensitive to drought, particularly during flowering (tasseling and silking) and grain-filling stages, where lack of water significantly reduces yield
  - Drought stress leads to reduced plant height, ear size, and kernel number
  - During flowering, water deficits result in silk drying, leading to significant yield reduction or failure
  - Maize prefers well-drained soils and is susceptible to waterlogging, especially during flowering and reproductive stages
  - Excessive moisture reduces growth, photosynthesis, and biomass production

- Impacts of Climate Change
  - Maize demands relatively optimum temperatures. Temperatures exceeding 35 degrees centigrade adversely affect the yield, as the crop is highly vulnerable to high temperature stress

  - Increased temperatures due to climate change reduce yields from 8 to 40 per cent
  - The reduction in the yield depends on the magnitude and duration of temperature
  - Climate change driven raising temperatures results in advancement of heading and maturity, leading to a significant shortening of the vegetative phase, reproductive phase, maturity phase, and the length of whole growing season
  - Model predictions indicate that by the end of 21 st Century maize yields may decrease by 10 to 20 per cent, even with optimum water availability, if GHG emissions are high
  - Studies at global level also indicate that under high GHG emission scenarios maize yields could decline by 24 per cent by 2030, necessitating adherence to limit the climate change driven global temperature increase below 1.5 degrees centigrade over base period
- Mitigation and adaptation strategies to Climate Change
  - Optimizing nitrogen fertilizer use: Excess nitrogen fertilizer contributes to more biomass production which is not desirable. Optimizing application rates and timing reduce emissions while maintaining yields. Adjusting fertilizer rates and timing can help meet the crop's nutrient needs under changing climate conditions and minimize nutrient losses due to water stress
  - Adjusting planting dates: Altering planting dates help maize avoid extreme temperatures and water stress during critical growth stages
  - Utilizing drought-resistant varieties: Developing and deploying maize varieties that can tolerate drought conditions is crucial for maintaining yields in drier regions
  - Implementing conservation agriculture practices: Techniques like no-till farming and crop rotation, can improve soil health, water retention, and reduce erosion making the crop more resilient to climate change impacts
  - Promoting integrated pest and disease management: Climate change can alter pest and disease pressure on maize crops. Integrated management strategies can help minimize yield losses due to these factors

11.3 Major Pulse Crops

11.3.1 Chickpea (*Cicer arietinum*)

- Chickpea is generally drought tolerant. However, extreme weather events like prolonged droughts and heat stress can significantly reduce yields. Chickpea also offers potential solutions to climate change, including their ability to fix nitrogen, reduce reliance on synthetic fertilizers, and potentially require less water than other crops
- Climatic Requirements
  - Temperature
    - Chickpea prefer daytime temperatures between 70-80°F and nighttime temperatures between 64-70°F
    - Chickpea is a cool season crop that does best in the temperature ranges mentioned above
  - Sunlight
    - Chickpea needs full sun exposure with at least 6-8 hours of direct light per day
  - Rainfall
    - While they can tolerate drier conditions, due to their deep tap root, they prefer well-distributed rainfall, as heavy rainfall can cause disease problems
  - Specific Regions
    - Chickpea is commonly grown in areas with Mediterranean and subtropical climates, including parts of South Asia, the Mediterranean region, and the Palouse region of the Pacific Northwest. It is also grown in the Northern Great Plains of North America
- Impacts of Climate Change
  - Drought: Drought is a major challenge, particularly in regions like sub-Saharan Africa and South Asia, where water scarcity is already a problem
  - Extreme Temperatures: Heat stress during flowering and pod setting can severely reduce yields

  - Yield Losses: Studies show that drought can lead to yield losses of up to 50% in some regions
- Mitigation and adaptation strategies to climate change
  - Nitrogen Fixation: Chickpeas, like other legumes, have a symbiotic relationship with nitrogen-fixing bacteria, reducing the need for synthetic nitrogen fertilizers, which are energy-intensive to produce
  - Water Efficiency: While drought can be a problem, chickpea is generally more drought-tolerant than some other crops, potentially requiring less irrigation and contributing to water conservation
  - Soil Health: Chickpea can improve soil health by enhancing soil structure, reducing erosion, and promoting biodiversity through their root systems

11.3.2 Pigeonpea (*Cajanus cajan*)

- Climatic Requirements
- Pigeonpea, a legume, thrives in tropical and subtropical climates with warm temperatures and moderate rainfall. It is notably drought-tolerant once established and can grow in a range of soil types, though well-drained, slightly acidic soils
  - Temperature
    - Optimal Range: Pigeonpea performs best within a temperature range of 20-30°C (68-86°F)
    - Heat Tolerance: It can tolerate higher temperatures, up to 35°C (95°F) or more, especially with adequate moisture
    - Frost Sensitivity: Pigeonpea is highly susceptible to frost damage and is typically grown as an annual in temperate zones
    - Pigeonpea is particularly vulnerable to heat stress, with even small temperature increases potentially reducing yields. High temperatures during flowering and pod development can negatively affect pollen viability, pollination, and seed set leading to reduced yields

- Rainfall
    - Drought Tolerance: Pigeonpea is well known for its drought tolerance and can survive on as little as 100 mm of rainfall
    - Ideal Range: However, it thrives with 600-1000 mm of rainfall annually
    - Waterlogging Sensitivity: Pigeonpea does not tolerate waterlogged conditions and requires well-drained soils
    - Changes in rainfall patterns, including both droughts and excessive rainfall, can also negatively impact pigeonpea production
- Day Length
    - Pigeonpea varieties can vary in their sensitivity to day length, with some requiring short days for flowering
    - It can be grown at elevations up to 2000 m

- Impacts of Climate Change
    - Climate change, particularly increased temperatures and altered rainfall patterns, is projected to negatively impact pigeonpea production. Pigeonpea, a crucial food security crop in many regions, is susceptible to heat stress and waterlogging, and its yield is sensitive to temperature and rainfall variations
    - Pigeonpea is particularly vulnerable to heat stress, with even small temperature increases potentially reducing yields. High temperatures during flowering and pod development can negatively affect pollen viability, pollination, and seed set, leading to reduced yields
    - Changes in rainfall patterns, including both droughts and excessive rainfall, can also negatively impact pigeonpea production
    - Climate change can also lead to shifts in pest and disease patterns further threatening pigeonpea crops. Some studies suggest that increased temperatures may lead to more generations of certain pests like <u>Helicoverpa armigera</u>, potentially increasing their impact on pigeonpea

    - Pigeonpea is a rainfed crop, and its yields are highly susceptible to water stress, which is likely to be exacerbated by climate change

- Mitigation and adaptation strategies to Climate change
    - Developing Heat-Resilient Varieties: Research focuses on developing pigeonpea varieties that are more tolerant of high temperatures and heat stress. This includes understanding how rising temperatures affect the root system of pigeonpea and optimizing root-soil interactions
    - Optimizing Planting Dates and Management Practices: Adjusting planting dates to avoid periods of high temperatures or water stress can help mitigate the negative impacts of climate change
    - Intercropping: Intercropping pigeon pea with other crops like maize can be a strategy to reduce yield losses under climate change scenarios, though the impact can vary depending on the specific varieties and climate models used

11.3.3 Soybean (*Glycine max.*)

- Climatic Requirements
- Soybean thrives in warm, temperate regions with adequate rainfall and sunlight
    - Temperature
        - Soybean prefer warm temperatures, with the most favorable range being between 20°C and 30°C
        - Temperatures below 10°C significantly delay development and reduce yield
        - While soybean tolerate higher temperatures, excessive heat (above 35°C) negatively impact growth and yield
        - The minimum temperature for germination and emergence is 10°C
    - Rainfall
        - Soybean need consistent moisture, but excessive rainfall can lead to fungal diseases and waterlogging
        - The ideal rainfall range is typically between 40cm and 60cm per year, with most of it needed during the growing season

    - Regions with well-drained soils are crucial to prevent waterlogging
  - Sunlight
    - Soybean require a good amount of sunlight for proper growth and development
    - While soybean grows in a variety of latitudes, it generally performs best in areas with long, hot summers
    - Photoperiod, or the length of daylight, also plays a role in soybean development, with different varieties adapted to different day lengths
- Impacts of Climate change
- Climate change is projected to have mixed effects on soybean yields, with some regions experiencing potential gains and others facing decreases. While warmer temperatures and longer growing seasons in some areas may lead to increased yields, overall, rising temperatures and increased frequency of extreme weather events are expected to negatively impact soybean production in many regions, particularly those experiencing drought
  - Positive Impacts
    - Extended Growing Seasons: Warmer temperatures and extended growing seasons in some northern regions (like the Dakotas and southern Canada) could allow for increased soybean production in areas where it was previously not possible
    - Enhanced Nitrogen Fixation: Increased vapor pressure deficit, a driver of atmospheric drying, benefit soybean's biological nitrogen fixation process, potentially boosting yields, especially in irrigated conditions
  - Negative Impacts
    - Increased Heat Stress: Extreme heat events can negatively impact soybean yields, causing reduced growth and seed set
    - Drought Impacts: Declines in precipitation and increased drought frequency, particularly east of the 100th meridian in the U.S., could significantly reduce soybean yields

- Mitigation and adaptation strategies to climate change
  - Water Management: Efficient water management practices, including harvesting and prioritizing water use are essential for mitigating the effects of drought
  - Genetic Improvement: Researching and developing soybean varieties that are more resilient to climate change, including heat and drought tolerance, is critical
  - Crop Rotation: Practicing crop rotation with crops like corn can help maintain soil health and improve soybean yields

11.4 Addressing Climate Change Challenges on Cereals and Pulse Crops

- To address climate change challenges on cereal and pulse crops, individuals, families, communities, governments, etc., have the best sciences available to plan for and implement climate-smart and resilient practices. Two critically important approaches which are available to managing climate resilience are mitigation and adaptation. These strategies are inter-related and must be pursued simultaneously as both can potentially improve resilience to the changing climate.

Chapter 12

Climate Change and Major Oilseeds and Commercial Crops, and Pests and Diseases

"Vedas revere Earth as Mother, emphasizing inter-connectedness and sustainable living. Oilseed crops, as Earth's bounty, foster ecological balance and support life's diversity. Nourishing the Earth sustains the atmosphere, embodying the Vedic wisdom of harmony and universal well-being"

12.1 Introduction to Major Oilseeds and Commercial Crops

- Oilseed crops and commercial crops like cotton, sugarcane, and cassava are essential agricultural commodities cultivated globally for a diverse range of uses, encompassing food, industrial applications, and bioenergy production
- Oilseed crops are a primary source of vegetable oils for cooking and food processing. The material left after oil extraction is rich in protein and used as animal feed and increasingly as functional ingredients in human food products

- Oilseed crops are a vital source of raw materials for various industrial uses, including the production of biodiesel, cosmetics, paints, lubricants, and pharmaceuticals
- Oilseeds contribute to sustainable agricultural practices by enriching soil fertility, providing crop rotation options, and reducing the reliance on synthetic fertilizers
- Cotton is the world's most widely used natural fiber, a cornerstone of the global textile industry
- Cotton is also used in a variety of industrial applications, including the production of fabrics for home furnishings and industrial products like tarpaulins and tents
- Cotton cultivation and processing provide income for millions of people worldwide and support entire industries, from farming to textile manufacturing
- Cottonseed, a byproduct of cotton processing is used to extract oil for cooking and industrial applications and serves as an ingredient in livestock feed.
- Sugarcane is a primary source for the production of sugar, a vital sweetener used globally in foods and beverages
- Sugarcane is a crucial feedstock for the production of ethanol, a renewable biofuel, and the byproducts of sugar processing, like bagasse and molasses, can also be converted into biofuel and electricity
- Sugarcane derivatives are used in a variety of industries, including the production of alcoholic beverages, plastics, paper, and chemicals
- Sugarcane cultivation and processing provide significant employment opportunities in rural areas and contribute to the economic development of many tropical countries
- Sugarcane has advantages like being effective against erosion and promoting energy self-sufficiency by using the entire plant for energy production

- Cassava is a staple food for over 800 million people, particularly in the tropics, providing a critical source of calories and energy
- It is drought-tolerant and thrives in poor soils, making it invaluable in areas facing climate change impacts or limited agricultural resources
- Due to its ability to be stored underground for long periods, cassava acts as a reliable famine reserve, offering a vital food source during times of scarcity

12.2 Major Oilseeds Crops

12.2.1 Groundnut (*Arachis hypogaea*)

- Groundnut, also known as peanut, thrives in tropical and subtropical regions worldwide, generally between 40°N and 40°S latitudes. Optimal growing conditions require specific temperature, rainfall, and soil characteristics
- Climatic Requirements
    - Temperature
        - Growing Season: Groundnut need a long, warm growing season with high temperatures, ideally between 20-30°C for optimal growth, flowering, and pod formation. A temperature of 20-30°C is considered optimum for growth and flower production
        - Germination: Soil temperatures around 27-30°C are best for rapid germination and seedling development. Minimum soil temperature for germination is about 18°C, with temperatures between 20-30°C resulting in 95% germination
        - Temperature Extremes: Groundnut is sensitive to cold and frost and cannot tolerate long periods of frost. Growth is inhibited above 35°C and stops entirely below 15°C
    - Rainfall
        - Amount: The total water requirement for a groundnut crop ranges from 500 to 700 mm over the growing period. While peanut is typically grown in areas with 600-1,500 mm of annual rainfall, 1,250 mm is considered optimal

        - Distribution: Rainfall needs to be well-distributed throughout the growing season, especially during critical stages like flowering, pegging, and pod formation
        - Sensitivity to Extremes: Excessive soil water hinders root growth and nitrogen fixation, while severe droughts lead to aflatoxin contamination, a toxic mold caused by fungi. Proper soil drainage is crucial to prevent waterlogging
    - Other Factors
        - Sunlight: Full sun exposure (at least eight hours daily) is best for optimal growth, although groundnut can tolerate partial shade
        - Altitude: Groundnut is grown at altitudes of up to 1,050 meters
- These climatic requirements are important for successful groundnut cultivation, particularly in the face of climate change challenges such as heat stress, drought, and erratic rainfall patterns
- Impacts of climate change
    - Climate change poses significant threats to the production of groundnut (peanut)
    - Drought: Groundnut is particularly vulnerable to drought stress, especially during critical growth stages like flowering, peg initiation, and pod filling. Drought significantly reduce pod yield (potentially by as much as 85%), diminish nutrient uptake, and result in smaller, lower-quality seeds with reduced viability. Prolonged drought, particularly during the pod-filling stage, has also been linked to increased levels of pre-harvest aflatoxin contamination, a harmful fungal toxin
    - Heat stress: Elevated temperatures, particularly above the optimal range of 20–35°C, adversely affect groundnut growth and development. Heat stress reduce pollen viability, impair pod setting, and decrease seed mass and overall yield. A combination of heat and drought stress exacerbate these effects, leading to substantial yield and quality losses

- Elevated CO_2: Increased atmospheric carbon dioxide (CO_2) have both positive and negative effects on groundnut. While elevated CO_2 enhance photosynthesis and lead to increased biomass and pod yield in C_3 plants like groundnut, it also decrease oil and protein content in the seeds
- Unpredictable rainfall patterns: Climate change lead to more extreme weather events, including floods, waterlogging, which also devastate groundnut crops. Excessive rainfall, especially near harvest time, combined with hot night temperatures, negatively impact seed quality
- Increased incidence of pests and diseases: Warmer temperatures and altered rainfall patterns create more favorable conditions for the survival and proliferation of pests and diseases, further threatening groundnut production

- Mitigation and adaptation strategies to climate change
  - By understanding the complex effects of climate change on groundnut and employing a combination of mitigation and adaptation strategies, it is possible to build resilience in groundnut production systems and ensure the long-term viability of this important crop
  - Breeding climate-resilient varieties: Developing groundnut cultivars with enhanced tolerance to drought, heat, and other climate-related stresses
  - Improving water management: Implementing efficient irrigation strategies, such as critical irrigation during drought-sensitive periods and using techniques like sprinklers help optimize water use
  - Sustainable soil management: Practices like cover cropping, organic mulching, and crop rotation improve soil health, moisture retention, and nutrient availability
  - Utilizing genomic tools: Advanced genomic tools, including genomic selection and marker-assisted breeding, are accelerating the development of climate-resilient groundnut varieties.

12.2.2 Rapeseed and Mustard (Brassica napus and Brassica rapa. and Brassica juncea)

- Rapeseed and mustard thrive in temperate climates, characterized by cool, moist conditions
- Climatic Requirements
  - Temperature
    - Optimal growth occurs within a range of approximately 10°C to 25°C
    - Temperatures below 3°C or above 35°C can hinder or halt crop growth
    - High temperatures, especially during flowering and seed development, negatively impact yield and quality
    - Temperatures exceeding 27°C lead to sterility and floral abortion
    - Frost should be avoided during the flowering stage as it significantly reduces yields
  - Rainfall
    - A fair supply of moisture is needed during the growing period
    - Precipitation requirement is generally around 350-450 mm
    - Well-distributed rainfall during the growing season, particularly during pod and seed development is essential
    - However, heavy or continuous rainfall during the flowering stage negatively affects the crop, potentially leading to fungal diseases
    - Rapeseed and mustard do not tolerate waterlogging conditions
  - Other Factors
    - Low humidity is beneficial, especially during the flowering stage

- Clear, dry weather with plenty of sunshine at maturity helps increase oil yield
- These are considered long-day crops in their photo-periodic response
- While some tolerance to drought exists, particularly in certain mustard types, prolonged dry, hot conditions can reduce yields

o Specific considerations for different varieties

- Toria (a type of rapeseed) is more sensitive to frost and cold, therefore it is usually sown earlier and harvested before frost sets in
- Yellow mustard varieties typically mature in 80 to 85 days, while brown and oriental types require 90 to 95 days

- Impacts of climate change
  - o Global warming and associated increases in aridity in some temperate regions where these crops are currently grown may impact their production
  - o Climate change is projected to negatively impact rapeseed and mustard yields, primarily due to increased temperatures
  - o Specifically, warmer temperatures shorten the crop's growth cycle, leading to reduced seed formation and lower overall yield
  - o Aridity and drought, also consequences of climate change, further decrease oil content and seed yield
  - o Studies indicate that higher temperatures significantly lower the yield of both rapeseed and mustard
  - o Increased temperatures lead to earlier flowering and maturity, shortening the period available for seed development and thus reducing yield
  - o Increased aridity and drought conditions reduce the oil concentration in rapeseed seeds

    - The effects of climate change on these crops will vary geographically, with some regions experiencing more severe impacts than others

- Mitigation and adaptation strategies to Climate Change
    - Measures like late sowing and growing long-duration varieties may help mitigate some of the negative impacts of climate change particularly for irrigated mustard

12.2.3 Sunflower (***Helianthus annuus***)

- Sunflowers thrive in climates with moderate rainfall, warm temperatures, and ample sunshine. It is susceptible to frost, especially during germination and early growth
- Climatic Requirements
    - Temperature
        - Optimal Range:18-25°C is ideal for growth
        - Germination: Soil temperature of 12°C - 20°C is optimal for germination
        - Frost Sensitivity: Sunflower is susceptible to frost damage, especially during the early stages of growth
    - Rainfall and other factors
        - Sunflower generally requires between 350 and 1500 mm of rainfall depending on the location and varieties
        - It can be grown under irrigation in arid climates
        - Sunflower need about 1 inch of water per week during the growing season, with adequate moisture for germination
        - Sunlight: Sunflower need plenty of sunshine to thrive
        - Altitude: Sunflower can be grown at various altitudes, but optimal growth is observed at lower altitudes
- Impacts of Climate Change
    - Climate change impacts on sunflower crop include both negative and positive. Higher temperatures, elevated CO_2, and altered precipitation patterns lead to reduced yields and oil content, especially in regions already experiencing water stress

- Negative Impacts
  - Heat stress: Increased temperatures, particularly during flowering, negatively impact yield and oil content, and alter fatty acid composition
  - Drought: Reduced rainfall and increased evaporation leads to water stress, especially during critical growth stages, resulting in lower yields
  - Pollination: Changes in pollinator activity due to climate change could affect sunflower pollination, which relies heavily on insects
- Positive Impacts
  - Increased CO_2: Elevated CO_2 can boost photosynthetic rates and potentially increase biomass and seed production, though this may bc offsct by othcr factors
  - New cultivation areas: Warmer temperatures and longer growing seasons in some northern regions could create new opportunities for sunflower cultivation

- Mitigation and adaptation strategies to Climate Change
  - Breeding for stress tolerance: Developing sunflower varieties that are more tolerant to heat and drought stress is crucial
  - Optimizing planting dates: Adjusting planting dates to avoid peak heat or drought stress help mitigate negative impacts
  - Water management: Implementing efficient irrigation strategies and utilizing water-saving techniques help manage water scarcity
  - Diversifying cropping systems: Integrating sunflowers into new regions and crop rotations help improve overall agricultural sustainability

12.3 Major Commercial Crops

12.3.1 Cotton (*Gossypium* hirsutum)

- Climatic Requirements
- Cotton thrives in warm and humid climates with ample sunlight and a lengthy frost-free period
  - o Temperature
    - Optimal Range: 20-37°C is ideal for most of the growing season
    - Heat Tolerance: While temperatures above 37°C can be detrimental, short periods (During a heatwave) up to 43°C are generally tolerated if humidity is high
    - Seed Germination: 15-35°C
    - Seedling Emergence: 18-30°C
    - Vegetative Growth: 20-30°C
    - Flowering and Boll Development: 25-35°C
    - Boll Opening and Maturation: 20-30°C
    - Frost-free Period: For long duration varieties Minimum 180 days, with 200 being preferable
  - o Rainfall
    - Ideal Rainfall: 850-1100 mm per year
    - Minimum Rainfall: At least 500 mm if evenly distributed
    - Rainfall Distribution: Rainfall should be well distributed, but heavy rainfall during flowering and boll development cause problems
  - o Other Climatic Factors
    - Humidity: Relative humidity between 50-60% is generally considered optimal, though high humidity can increase disease risk
    - Sunlight: Cotton requires ample sunlight throughout the growing season
    - Wind: Moderate winds can be beneficial for pollination and humidity regulation, but strong winds can damage bolls or uproot plants

- Impacts of Climate Change
  - Climate change presents significant challenges to cotton production globally. Rising temperatures, increased frequency of extreme weather events, and altered precipitation patterns are expected to negatively impact cotton yields and quality. However, some studies suggest that increased CO_2 levels and temperature increases can also have positive effects on cotton growth in certain regions
  - Negative Impacts
    - Temperature Stress: Cotton is particularly vulnerable to temperature fluctuations, especially during flowering and boll formation. Rising temperatures can lead to heat stress, causing rapid flowering, reduced pollen viability, and increased flower abortion, ultimately decreasing the number of bolls and yield
    - Water Stress: Climate change is projected to intensify drought and alter rainfall patterns, leading to water scarcity in many cotton-producing regions. In certain regions cotton is a water-intensive crop, and water stress can significantly reduce yields
    - Extreme Weather Events: Increased frequency and intensity of extreme weather events like floods, heatwaves, and strong winds can directly damage cotton crops, especially during vulnerable stages of development. For example, extreme rainfall can lead to water logging, while strong winds can damage bolls and spoil the cotton fiber
    - Pest and Disease Outbreaks: Climate change can alter pest and disease cycles, leading to increased outbreaks and potentially impacting cotton production
  - Positive Impacts
    - CO_2 Fertilization Effect: Increased atmospheric CO_2 concentrations can enhance photosynthesis and biomass production in cotton, potentially boosting yields
    - Extended Growing Season: In some regions, higher temperatures may lead to a longer growing season, potentially increasing yields

- Mitigation and adaptation strategies to Climate Change
  - Developing Stress-Tolerant Varieties: Breeding cotton varieties that are more tolerant to heat, drought, and other climate-related stresses is crucial
  - Improving Water Management: Implementing efficient irrigation techniques and developing water-efficient cotton varieties can help mitigate the impacts of water scarcity
  - Adapting Farming Practices: Adjusting planting dates, optimizing fertilizer use, and adopting conservation agriculture practices can help cotton farmers adapt to changing climate conditions
  - Utilizing Climate Information: Accessing and utilizing climate information and forecasts can help farmers make informed decisions about planting, irrigation, and pest management
  - Climate change poses significant risks to global cotton production, but climate resilient agricultural practices, adaptation strategies and technological advancements can help mitigate these impacts and ensure sustainable cotton production in the future

12.3.2 Sugarcane (*Saccharum* officinarum)

- Climatic Requirements
- Sugarcane thrives in warm, tropical, and subtropical climates with ample sunlight and rainfall. It requires a long growing season, ideally 9-18 months. Frost is detrimental, and dry, cool, and sunny conditions are preferred during the ripening and harvesting phases
  - Temperature
    - Optimal Temperature: A mean temperature of 28-30°C is ideal for growth
    - Sprouting: 32-38°C is optimal for bud sprouting
    - Growth: 20-30°C is generally considered the optimum temperature range for growth
    - Ripening: Lower temperatures (20-10°C) are favorable for sugar accumulation during the ripening phase

    - Frost: Frost is extremely damaging and should be avoided
    - High Temperatures: Temperatures above 50°C can halt growth
  - Rainfall
    - Ideal Rainfall: A total rainfall of 1000-1500 mm well distributed throughout the year is suitable
    - Dry Period for Ripening: A dry period during the ripening phase is beneficial for sugar accumulation
    - Irrigation: Irrigation is necessary in areas with lower rainfall or during dry spells
    - Water Stress: High water stress can negatively impact the crop, particularly during germination and flowering
  - Humidity
    - High humidity (80-85%) is beneficial during the grand growth period for cane elongation. Moderate humidity (45-65%) with limited water supply is favorable during ripening
  - Sunshine
    - Sugarcane is sun loving crop. Therefore, it requires plenty of sunlight for optimal growth and sugar production. Bright sunshine of 7-9 hours of sunshine per day are ideal. Greater solar radiation results in higher sugar yields
  - Altitude
    - Sugarcane can be grown at altitudes up to 1000 meters above sea level
- Impacts of Climate Change
  - Climate change significantly impacts sugarcane production, affecting yields and supply chains. Increased temperatures, unpredictable rainfall, and extreme weather events like droughts and floods can negatively impact sugarcane growth and sugar content. These changes also increase the risk of pests and diseases, further threatening production

- Yield and Quality: Climate change-related factors like temperature, rainfall, and extreme weather events directly influence sugarcane yield and sugar content. Increased temperatures can reduce yields, while drought or excessive rainfall can negatively impact growth and sugar accumulation
- Supply Chains: Changes in sugarcane production due to climate change can disrupt entire supply chains, impacting the availability and price of sugar and related products
- Pests and Diseases: Climate change can alter pest and disease patterns, potentially increasing their impact on sugarcane crops
- Water Availability: Changes in rainfall patterns can lead to water scarcity in some regions, affecting irrigation and sugarcane growth
- Soil Salinity: Rising sea levels can lead to salinization of sugarcane fields, making them unsuitable for cultivation

- Mitigation and adaptation strategies to Climate Change
  - Breeding for Resilience: Developing sugarcane varieties that are tolerant to heat, drought, and diseases is crucial for adapting to climate change
  - Improved Cropping Systems: Implementing sustainable agricultural practices, such as water management techniques and integrated pest management, can help mitigate the negative impacts
  - Climate-Smart Technologies: Utilizing technologies like precision agriculture and climate-resilient infrastructure can improve sugarcane production and reduce its environmental footprint
  - Policy and Planning: Governments and organizations need to develop policies and strategies to support sugarcane farmers in adapting to climate change and promoting sustainable practices

12.3.3 Cassava (*Manihot* esculenta)

- Climatic Requirements
- Cassava is a versatile crop well-suited to tropical and subtropical regions with warm temperatures, adequate rainfall, and suitable soil conditions. It thrives in warm, humid, and frost-free climates. The plant is sensitive to photoperiod, with longer photoperiods inhibiting starch storage
  - Temperature
    - Optimal: 25-29°C
    - Critical Minimum: Growth stops at or below 10°C
    - Tolerance: 15-35°C
    - Ideal Range: 20-27°C
  - "Rainfall"
    - Optimal: 1000-1500mm annually
    - Dry Season Tolerance: Can tolerate prolonged periods of drought
  - Other Factors
    - Photoperiod: 12-hour photoperiod is optimal; longer periods can inhibit starch storage
    - Altitude: Typically grown in tropical lowlands, but can be found up to 1500 m
- Impacts of Climate Change
  - Cassava, a staple crop in many regions, shows resilience to climate change. While some areas may experience decreased suitability for cassava due to climate change, its ability to thrive in harsh conditions like drought and poor soils makes it a valuable crop for food security, particularly in Sub-Saharan Africa
  - Increased Temperatures: Rising temperatures can negatively affect cassava yields, especially in areas already experiencing high temperatures

- o Changing Rainfall Patterns: Unpredictable rainfall, including droughts and floods, can damage cassava crops and affect yields
- o Pests and Diseases: Climate change can exacerbate pest and disease outbreaks, impacting cassava production.

- Cassava's Role as Climate Resilience crop under Climate Change
  - o Drought Tolerance: Cassava is known for its ability to withstand drought conditions, making it a reliable food source when other crops fail
  - o Poor Soil Tolerance: It can grow in nutrient-poor soils, which are common in many regions affected by climate change
  - o Flexibility in Planting and Harvesting: Cassava can be planted and harvested at different times, offering flexibility in adapting to unpredictable weather patterns
  - o Potential for Increased Yields: While some studies suggest potential decreases in cassava yields due to climate change, others indicate that it may be able to maintain or even increased yields in certain regions, especially with improved varieties and farming practices
  - o Food Security Crop: Its ability to withstand harsh conditions makes it a valuable crop for ensuring food security in vulnerable regions. In many parts of the world, particularly in Africa, cassava is a crucial food security crop, providing a reliable source of calories and nutrients for millions
  - o Cassava can be a key part of climate change adaptation strategies for farmers, helping them to mitigate the negative impacts of climate change on food production. Cassava can also be used for various industrial purposes, such as biofuel production, which can further contribute to climate change mitigation efforts

12.4 Climate Change and Pests and Diseases on Agricultural Crops

- Research results indicate that climate change affects agriculture because of increased temperatures, changes in rainfall patterns and increased frequency of extreme events, which could cause changes in pest and disease ecology, ecological disruption in agricultural areas, socio-economic shifts, and in land-use practices. These changes can also create more favorable conditions for pest and pathogen development and spread, leading to potential yield losses and reduced grain quality

12.4.1 Generalized Impacts of Climate Change on Crop Pests

- Warmer temperatures can expand the range of many pests, allowing them to survive in regions previously too cold for them. For example, some pests require specific temperature ranges to become active and cause damage to crops
- Warmer temperatures can allow pests to complete more generations per year, leading to increased damage to crops
- Warmer temperatures and longer growing seasons can lead to increased pest reproduction rates, allowing them to complete more generations per year and potentially causing larger outbreaks
- Warmer temperatures can cause pests to expand their geographical range, moving into areas where they were previously absent or less prevalent
- Warmer temperatures and changes in precipitation can lead to increased pest populations and more frequent outbreaks, affecting the growth and yield of pulse crops
- Changes in rainfall patterns, including increased frequency of heavy rainfall and prolonged droughts, can directly influence the spread of pests. Heavy rainfall can lead to flooding and increased humidity, while drought can stress plants, making them more susceptible to pest attacks

- Events like storms and floods can physically disperse pests, disrupt ecosystems, and weaken plants, making them more vulnerable to outbreaks
- Climate change can disrupt the balance between pests and their natural enemies, potentially leading to increased populations
- Some pests and insects may exhibit increased feeding rates and voracity as temperatures rise

12.4.2 Generalized Impacts of Climate Change on Crop Diseases

- The survival rate of pathogens in winter or summer could vary with increase in surface temperature. Higher temperatures in winter will not only result in higher pathogen survival rates but also lead to extension of cropping area, which could provide more host plants for pathogens
- Thus, the overall impact of climate change is likely to be an enlargement of the source, population, and size of pathogens
- Damage from disease may be more serious because heat-stress conditions will weaken the disease–resistance of host plants and provide pathogens with more favorable growth conditions
- Warmer temperatures, increased humidity, and altered rainfall patterns can create favorable conditions for pathogen survival, reproduction, and dispersal, leading to more frequent and intense disease outbreaks
- Changes in temperature and humidity can affect the life cycles of pathogens, influencing their ability to survive, reproduce, and spread
- Changes in temperature and precipitation can affect the survival and distribution of vectors, like insects that transmit plant diseases, potentially leading to the spread of new diseases
- Many fungal pathogens thrive in warm and humid conditions making them more prevalent with climate change

- Climate change can alter the timing and geographic distribution of plant diseases, affecting crop production in new regions or during different times of the year
- Climate change can affect the efficacy of fungicides and other disease control measures, making disease management more challenging
- Climate change can alter the distribution and severity of plant diseases, potentially leading to new disease outbreaks in regions where they were previously not prevalent
- Climate change can create conditions that favor the emergence of new diseases or the reemergence of previously controlled diseases

12.4.3 Other Impacts of Climate Change on Crop Pests and Diseases

- Elevated CO_2 levels can weaken plant defenses, making them more vulnerable to pathogens
- Increased CO_2 concentrations can affect pathogen sporulation and disease development, with potential consequences for various pulse crops
- Higher CO_2 levels can lead to lower concentrations of essential nutrients like iron and protein in pulses like soybean, beans, and other legumes, impacting food security and nutritional value, especially in vulnerable populations
- Some pathogens may become more aggressive or virulent under changing climatic conditions, causing more severe crop damage

12.4.4 Examples of Climate Change Impacts on Pests and Diseases

- Cereals, Pulses, and other crops
    - Wheat rusts: Rising temperatures are allowing wheat rusts to spread to new regions and evolve into more aggressive forms. Rising temperatures are also driving the evolution and spread of wheat rusts, such as stem rust, which can cause significant yield losses

- Aphids: Elevated temperatures and CO_2 levels can increase aphid populations and enhance their ability to transmit viruses, impacting cereal crops like wheat
- Locusts: Climate change, particularly increased temperatures and altered rainfall patterns, can create favorable conditions for locust outbreaks, potentially causing widespread crop damage
- Rice blast: Warmer temperatures and increased humidity can promote the development and spread of rice blast, a major disease of rice crops
- Soybean Rust: Rising temperatures and humidity are contributing to the spread of soybean rust, a fungal disease, with studies showing significant farmer impacts in places like Malawi
- Pea Aphids: Climate change scenarios, particularly those with increased temperatures and moisture, are predicted to influence the global expansion of pea aphids, a significant pest of peas
- Soil-borne pathogens: Changes in soil pH due to climate-related shifts in land use and rainfall patterns can affect the distribution and abundance of soil-borne pathogens like Phytophthora and Pythium, potentially leading to increased disease outbreaks
- Charcoal Rot: This fungal disease, which affects soybean and other crops, is expected to be more prevalent under warmer and drier conditions, potentially impacting yield
- Asian Soybean Rust: Warmer temperatures and increased humidity can create favorable conditions for the spread of Asian soybean rust, a major disease affecting soybean crop

- Oilseed Crops
  - Phoma Stem Canker: Increased mean winter temperatures have been linked to enhanced infection rates of phoma stem canker in oilseed rape
  - Sclerotinia Stem Rot: Changes in temperature and humidity can influence the severity of Sclerotinia stem rot in oilseed crops

- Potato Diseases: Changes in temperature and rainfall patterns are impacting the severity and distribution of various potato diseases, including late blight and bacterial infections
- Cotton Bollworm: Increased winter and spring minimum temperatures favor Cotton Bollworm outbreaks, particularly in temperate regions
- Cotton Aphids: Warmer temperatures can increase the reproduction rate and lifespan of cotton aphids, leading to larger populations and more damage
- Root Rot: Increased rainfall can lead to more root rot in cotton, as waterlogged soil favors the development of pathogens
- Yellow Sugarcane Aphids (Sipha flava): Their populations are increasing due to climate change, impacting sugarcane productivity
- White Scales (Pseudaulacaspis pentagona): These are also becoming more prevalent due to climate change
- White Grubs (Phyllophaga sp.): Their populations are also impacted by climate change
- Plassey borers: Climate change can affect the behavior and spread of these pests
- Top borers: These insects are also impacted by climate change, potentially leading to increased damage to sugarcane

- Conclusions
  - Given the complex interactions between climate change and plant pests and diseases, developing effective adaptation and mitigation strategies is crucial for protecting crop production. This includes: Developing disease-resistant crop varieties, implementing sustainable pest management practices, improving irrigation and water management, developing early warning systems for pest and disease outbreaks, and promoting climate-smart agricultural practices

12.5 Addressing the issues of Climate Change on Oilseeds and Pulse Crops

- To address climate change challenges, individuals, families, communities, governments, organizations, etc., have the best sciences available to plan for and implement climate-smart and resilient practices. Two critically important approaches which are available to managing climate resilience are mitigation and adaptation. These strategies are inter-related and must be pursued simultaneously as both can potentially improve resilience to the changing climate.

Chapter 13

Climate Change and Major Horticultural Crops

"According to Vedic wisdom, Horticulture plantations and fruit crops are not merely sources of sustenance but are revered as embodiments of nature's bounty and an integral part of maintaining the Earth's delicate balance. The Vedas emphasize the interconnectedness of all life and the importance of responsible stewardship of the environment, considering the Earth as mother and humanity as her children. The planting and tending of trees and plants are considered a sacred duty (dharma) that contributes to a harmonious coexistence with nature and the well-being of all beings"

13.1 Introduction to Major Horticultural Crops

- Horticultural crops, encompassing fruits, plantation crops, vegetables, etc., are cultivated for a multitude of reasons spanning nutritional, economic, and environmental benefits
- These crops provide a wide variety of essential nutrients, including vitamins, minerals, and fiber crucial for a healthy and balanced diet
- Diets rich in fruits and vegetables can significantly reduce the risk of nutrient-related disorders and improve food security for individuals and nations

- Most fruits and vegetables are naturally low in calories and fat, but high in fiber and water content, helping individuals feel full and satisfied while managing weight
- Horticultural crops, often considered high-value crops, can generate significantly more income per unit of land compared to traditional field crops
- The horticulture sector creates extensive employment opportunities, from cultivation and processing to distribution and retail, benefiting rural and urban communities alike
- Horticulture can be a powerful engine for agricultural and economic diversification, especially for smallholder farmers and marginalized communities
- Countries with suitable climates and expertise in horticulture can capitalize on their resources by exporting high-quality produce to meet international demands, boosting national economies
- Horticulture provides opportunities to utilize small spaces and even marginal lands productively, fostering rural entrepreneurship and creating livelihood options
- Horticulture promotes biodiversity by providing habitats for a diverse range of plant and animal species, including pollinators and beneficial insects
- Plants and green spaces absorb carbon dioxide, release oxygen, and can reduce the urban heat island effect, mitigating the impacts of climate change
- Sustainable horticultural practices like cover cropping, crop rotation, and composting improve soil health, reduce erosion, and enhance water retention
- Techniques like drip irrigation and rainwater harvesting optimize water use, reducing waste and addressing water scarcity concerns
- Horticultural plants help absorb pollutants, control dust, and filter rainwater, improving air and water quality

13.2 Major Fruit Crops of the World

13.2.1 Mango (*Mangifera indica*)

- Climatic Requirements
- Mango trees thrive in warm, tropical, and subtropical climates
    - Temperature
        - 20°C to 30°C (75°F to 86°F) for optimal growth and fruit production.
        - Temperatures below 10°C (50°F) can damage or kill young trees, and prolonged exposure to low temperatures can severely defoliate mature trees
        - Established trees can tolerate high temperatures, even up to 48°C (118°F), but young trees need protection from extreme heat until they establish a strong root system
        - Dry, warm conditions are ideal for flowering, while rain during flowering can hinder pollination
    - Sunshine
        - Mango trees need ample sunlight (8 hours per day) for proper growth and fruit production
    - Rainfall
        - While mangoes can tolerate a range of rainfall, they prefer a distinct dry season for flowering and a wet season for growth
    - Humidity
        - High humidity during flowering can hinder fruit set, and high winds can cause flower and fruit shedding
- Impacts of Climate change
    - Climate change is significantly impacting mango production worldwide. Warmer temperatures, increased rainfall variability, and extreme weather events like heatwaves and droughts are disrupting mango growth cycles, affecting flowering, fruit development, and overall yield. This leads to reduced fruit quality, quantity, and potential economic losses for farmers

- Mangoes are highly sensitive to temperature fluctuations, particularly during flowering. Climate change-induced temperature increases can disrupt the flowering process, leading to reduced fruit set or even complete failure to fruit
- Extreme temperatures, especially heatwaves, can cause fruit drop, premature ripening, and the development of spongy tissue within the fruit, impacting its quality and market value
- Drought and water scarcity, often exacerbated by climate change, can severely stress mango trees, leading to reduced fruit size and quality
- Changes in temperature and humidity can affect the proliferation of pests and diseases, further impacting mango production
- While some areas may experience increased suitability for mango cultivation due to warmer temperatures, other established mango-producing regions may face challenges due to increased heat, drought, and other climate-related issues
- Specific Examples
  - Italy: While some areas in Italy are seeing increased mango cultivation due to climate change, others are experiencing challenges like heatwaves and extreme weather events impacting ability of mango trees to flower and fruit
  - Mexico: Drought and water scarcity have led to widespread problems with mango fruit quality and agricultural productivity
  - Peru: Unusually warm temperatures, compounded by El Niño effects, contributed to a potentially disastrous mango season

- Mitigation and adaptation strategies to Climate Change
  - Breeding and selecting mango varieties that are more tolerant to heat, drought, and other climate-related stresses is crucial

  - Implementing efficient irrigation systems like drip or sprinkler irrigation can help manage water resources effectively and mitigate drought impacts
  - Integrating mango trees with other crops or using shade management techniques can help moderate temperature extremes
  - Utilizing cover crops, intercropping, and other sustainable agricultural practices can improve soil health and resilience to climate change
  - Establishing systems to monitor weather patterns and provide timely warnings to farmers about potential climate-related risks can help them prepare and adapt

13.2.2 Orange (*Citrus sinensis*)

- Climatic Requirements
- Orange trees thrive in warm, subtropical or tropical climates
  - Temperature
    - Orange trees prefer warm weather with an optimal range between 60°F and 90°F (15°C and 32°C)
    - They can tolerate temperatures as high as 95°F (35°C)
    - However, temperatures below 29.5°F (-1.4°C) can cause damage to green oranges, and temperatures below 28°F (-2.2°C) can damage ripe oranges
  - Sunshine
    - Orange trees need a lot of sunshine for optimal growth and fruit production
  - Rainfall
    - They require a moderate amount of water, with adequate rainfall or supplemental irrigation if rainfall is insufficient
  - Frost
    - Orange trees are sensitive to frost and cold temperatures. Temperatures below freezing can damage the trees and the fruit. They are dormant in the winter and more tolerant of lower temperatures during this period, but prolonged exposure to freezing temperatures can be harmful

- Humidity
    - High humidity can increase the risk of diseases, while dry conditions are generally favorable
- Wind
    - Strong winds can cause fruit loss and damage the trees due to vibration

- Impacts of Climate Change
    - Global climate change significantly impacts orange production through increased temperatures, altered precipitation patterns, and more frequent extreme weather events. These changes can lead to reduced fruit quality, lower yields, and increased susceptibility to diseases
    - Impact on Fruit Quality
        - Higher temperatures and water stress can lead to oranges with less juice and lower sugar content
        - Extreme heat can cause fruit to develop blemishes, cracks, and other imperfections making them less appealing for fresh consumption
        - Warmer temperatures can delay fruit ripening, affect the timing of harvests and potentially impacting fruit quality
    - Impact on Yield and Production
        - Drought and high temperatures can significantly reduce fruit set (the number of oranges that develop on a tree), leading to lower yields
        - Warmer and wetter conditions can create an environment more favorable for the spread of citrus diseases like citrus greening further impacting yields and tree health
        - Hurricanes, freezes, and other extreme weather events, which are becoming more frequent and intense due to climate change, can cause significant damage to orange groves leading to crop losses

- Specific Examples
  - Florida: Hurricane Ian and Nicole, have severely impacted Florida's orange production, with some growers abandoning their land
  - Brazil: Recent heatwaves in Brazil, a major orange juice producer, have stressed orange trees and reduced juice availability
  - Iran: Climate change impacts on citrus crops, including oranges, in Iran can affect farmers, consumers, and the national economy

- Mitigation and Adaptation Strategies to Climate Change
  - Developing more climate-resilient orange varieties
  - Improving irrigation and water management
  - Combating citrus greening
  - Precision agriculture

13.2.3 Banana (*Musa paradisiaca*)

- Climatic Requirements
- Bananas thrive in warm, humid, and tropical or subtropical climates
  - Temperature
    - Bananas prefer warm temperatures, with optimal growth occurring between 20°C and 30°C (65°F and 86°F)
    - Growth slows significantly below 14°C (57°F)
  - Humidity
    - High humidity (50% or higher) is essential for healthy growth and fruit development
  - Sunshine
    - Bananas need plenty of sunlight, ideally around 12 hours of bright solar radiation per day

- Rainfall
  - While bananas require a lot of water, they are also susceptible to diseases like Panama wilt if their roots are waterlogged
  - Adequate rainfall, well distributed throughout the year is crucial
  - Areas with 2000 to 2500 mm of rainfall per year are suitable, but bananas can also be cultivated in areas with less rainfall with irrigation
- Altitude
  - Bananas can be grown from sea level up to 1200 meters in altitude
- Wind
  - Bananas are susceptible to damage from strong winds, so they should be protected from strong winds or cyclones
- Latitude
  - While most banana-growing regions are between the equator and 20°N and 20°S, some commercially viable plantings exist at higher latitudes (up to 30°N and 30°S) in subtropical regions with mild climates

- Impacts of Climate change
  - Climate change is significantly impacting banana production globally, with rising temperatures and shifting weather patterns threatening both yields and the suitability of growing regions. Specifically, rising temperatures can exceed the optimal growth range for bananas, leading to decreased productivity, while changes in rainfall patterns can exacerbate fungal diseases like Black Sigatoka and Panama disease. These factors are causing a decline in suitable growing areas and potentially disrupting the global banana supply chain
  - Banana varieties are sensitive to temperature fluctuations, and warming trends can push temperatures beyond their optimal growing range, negatively impacting yield

    - Some regions currently ideal for banana cultivation may become too hot, while others, further from established infrastructure, may become more suitable, requiring adaptation strategies
    - Changes in rainfall patterns, including increased drought and flooding, can stress banana plants and increase the risk of fungal diseases like Black Sigatoka, which thrives in wet conditions
    - Warmer, wetter conditions can also favor the spread of Panama disease (Fusarium wilt), a soilborne disease that can devastate banana plantations
    - Socioeconomic Impacts
        - Bananas are a major export crop for many countries, and disruptions to production can have significant economic consequences for farmers and communities that rely on banana revenues
        - Bananas are a staple food for millions of people, and reduced yields could lead to food shortages and increased prices
- Mitigation and Adaptation Strategies to Climate Change
    - Breeding bananas that can withstand higher temperatures is one approach to mitigating the effects of climate change
    - Implementing sustainable farming techniques, such as water conservation and integrated pest management can help improve resilience
    - In some cases, farmers may need to relocate to areas where climate conditions are more favorable for banana cultivation

13.2.4 Grape (*Vitis vinifera*)

- Climatic Requirements
- Grapes, particularly those used for winemaking, have specific climatic requirements for successful cultivation. Ideal conditions include a suitable temperature range, sufficient sunlight, and appropriate rainfall or irrigation. Temperature is

crucial, with different varieties thriving in cool, moderate, or warm climates. Sunlight is essential for photosynthesis and fruit development, and excessive moisture during ripening can lead to issues like diluted flavors and fungal diseases

- Temperature
  - Generally, grapes prefer an average growing season temperature (spring through fall) between 61.7°F (16.5°C) and 69.8°F (21°C), with specific varieties having narrower ranges
  - Grape varieties like Pinot Noir, Riesling, and Sauvignon Blanc thrive in cooler climates, generally below 61.7°F (16.5°C)
  - Varieties like Chardonnay, Merlot, and Cabernet Franc are well-suited to moderate climates, typically between 61.7°F (16.5°C) and 65.3°F (18.5°C)
  - Warmer climates, generally above 65.3°F (18.5°C), can support grapes like Zinfandel, Cabernet Sauvignon, and Syrah
  - Some grape varieties are more susceptible to winter injury due to low temperatures, which can damage buds, vascular tissues, or even lead to vine death
  - Grape varieties also differ in their tolerance to heat. As the climate warms, there will be a greater need to cultivate heat-tolerant varieties
- Sunlight
  - Grapevines require adequate sunlight for photosynthesis, the process by which they convert light energy into chemical energy for growth
  - Sunlight is particularly important during fruit setting and ripening, as it influences sugar accumulation and other factors affecting wine quality
  - Sunlight can also help reduce fungal diseases by drying out foliage and fruit

- Rainfall
  - Grapevines have specific water needs, and irrigation may be necessary in arid or semi-arid regions
  - High humidity and rainfall during the ripening period can lead to berry swelling, diluted flavors, and fungal diseases like botrytis bunch rot
  - Canopy management techniques, such as pruning and leaf removal can help manage airflow and reduce humidity around the fruit, minimizing disease risk
- Other Factors
  - Sloping vineyards can offer advantages in terms of drainage and sun exposure, with southern slopes in the Northern Hemisphere often preferred for maximizing heat accumulation
  - Large bodies of water can moderate temperature fluctuations, extending the growing season and potentially preventing vine damage

- Impacts of Climate Change
  - Climate change is impacting grape growing regions, with rising temperatures and altered rainfall patterns affecting suitability for different varieties. Climate change is significantly impacting grape growing and wine production globally. Rising temperatures, altered precipitation patterns, and increased frequency of extreme weather events are all affecting grapevines, their fruit, and the resulting wine quality. These changes are leading to shifts in suitable growing regions, disruptions in ripening processes, and challenges in maintaining the desired characteristics of wine
  - Warmer temperatures are pushing optimal grape-growing areas towards higher latitudes and altitudes. Some regions may become unsuitable for certain grape varieties, while others may see increased suitability
  - Warmer temperatures can cause grapes to ripen earlier, potentially leading to grapes being harvested at higher sugar levels and lower acidity. This can affect the overall balance and flavor profile of the wine

- Extreme heat can negatively impact grapevine physiology, reducing photosynthesis, slowing sugar accumulation, and potentially leading to sunburn or shriveling of the fruit
- Increased drought frequency and intensity can reduce yields and impact fruit quality. Water scarcity can also lead to conflicts over water resources, impacting irrigation practices
- Warmer temperatures and altered rainfall patterns can favor the spread of certain pests and diseases, posing additional challenges for grape growers, according to a review of studies
- The chemical composition of grapes is directly affected by climate change, influencing the aroma, color, and taste of the resulting wine

- Mitigation and Adaptation Strategies to Climate Change
    - Choosing grape varieties that are more tolerant to heat and drought
    - Modifying training systems to optimize sunlight exposure and water use
    - Implementing efficient irrigation strategies to manage water resources
    - Adjusting canopy density to regulate fruit exposure to sunlight and temperature
    - Some wine regions are exploring the possibility of moving vineyards to higher altitudes or different geographic locations that are more suitable for grape growing under changing climate conditions
    - Winemakers are adapting their practices to address changes in grape composition, including acid management, and fermentation techniques

13.2.5 Apple (*Malus domestica*)

- Apple trees generally require a temperate climate with distinct seasons, including cold winters and mild to warm summers
- Climatic Requirements
  - Temperature
    - Apple trees need a period of cold temperatures (below 45°F or 7°C) during dormancy, typically between 500 and 1000 chilling hours, to ensure proper bud break and flowering in the spring
    - While they need cold during dormancy, apple trees also need warm, but not excessively hot, temperatures during the growing season for fruit development. Ideal summer temperatures range from 65°F to 75°F (18°C to 24°C)
    - Temperatures consistently above 90°F (32°C) can stress the trees, reduce fruit quality, and make them more susceptible to diseases
  - Rainfall
    - Apple trees prefer moderate rainfall 750-1000 mm annually
  - Sunlight
    - Adequate sunlight is necessary for proper fruit development, influencing color, and overall fruit quality
  - Specific Considerations
    - Different apple varieties have varying chilling hour requirements. Some varieties may need more chilling hours than others
    - Areas near large bodies of water like lakes, can benefit apple orchards by moderating temperatures, reducing frost damage in the spring, and preventing extreme heat in the summer
    - While traditionally grown at certain altitudes (above 1600 ft or 487m), apple cultivation has expanded to various altitudes due to improved tree resilience and careful variety selection

- Impacts of Climate change
    - Global climate change is impacting apple production, primarily through alterations in temperature and precipitation patterns. These changes can lead to earlier blossoming, affecting the quantity and quality of the harvest as well as increasing the risk of frost damage and pest seasons
    - Altered Phenology (timing of events)
        - Warmer temperatures cause apple trees to blossom earlier in the spring
        - Earlier blooming makes apples more vulnerable to late spring frosts, which can damage blossoms and significantly reduce yields
        - Changes in bloom time can also lead to a mismatch between apple blooming and the availability of pollinators, impacting fruit set
    - Impact on Fruit Quality
        - Some apple varieties, like McIntosh, may not develop their characteristic red color or crispness due to warmer temperatures
        - Higher temperatures and increased sunlight can lead to sunburn on apples, reducing their marketability
        - Climate change can also affect the thickness and quality of the apple skin, making it more susceptible to damage
    - Changes in Yield
        - Some studies suggest that in certain scenarios, climate change might initially increase apple yields due to advanced flowering and improved growing conditions
        - However, other studies indicate that the overall impact on apple yields is negative, with a projected decrease in yields as climate change severity increases
        - The effects of climate change on apple production can vary significantly depending on the specific location, apple variety, and local climate conditions

- Other Impacts
  - Warmer temperatures can lengthen the pest and disease season, potentially increasing the need for pesticide use
  - Changes in precipitation patterns, such as increased drought or intense rainfall, can lead to water stress in apple trees, impacting fruit quality and yield
  - Climate change can also affect ecosystem services provided by apple orchards, such as carbon sequestration and water availability
- Mitigation and Adaptation Strategies to Climate Change
  - One way to adapt is to breed and cultivate apple varieties that are more tolerant to heat and drought
  - Shade cloths can be used to protect apples from sunburn during extreme heat events
  - Adjusting the layout of orchards to minimize sun exposure can also help reduce sunburn
  - In some cases, delaying flowering can help avoid late spring frosts

13.3 Major Plantation Crops of the World

13.3.1 Coconut (*Cocos nucifera*)

- Climatic Requirements
- Coconut is predominantly a tropical and subtropical crop
  - Temperature
    - Coconut palms flourish in warm, humid climates with average annual temperatures around 27°C (81°F)
    - The optimal temperature range for the coconut palm to thrive and produce fruit lies between 20°C and 30°C. Prolonged heat waves above this range negatively impact growth and productivity

        - Coconut palms are intolerant to cold weather and are susceptible to damage if temperatures drop below 4°C (39°F)
    - Rainfall
        - Coconut palms require consistent and well-distributed rainfall throughout the year
        - An annual rainfall of 1500 to 2500 mm (59 to 98 inches) is preferred, with no prolonged dry periods
        - Coconut palms are sensitive to drought. In such conditions, they exhibit signs of stress like leaf desiccation, premature fruit drop, and stunted growth. Irrigation becomes crucial in areas with insufficient rainfall or uneven distribution
        - Interestingly, coconut palms can tolerate temporary flooding and even benefit from high winds, which increase transpiration and nutrient uptake
    - Humidity
        - Coconut palms require high humidity levels, ideally between 70% and 80%, for optimal growth and development
        - Coconut palms are rarely found in areas with low humidity, even if temperatures are sufficiently high
    - Sunlight
        - Coconut palms are sunworshippers and require plenty of sunlight, ideally at least 6-8 hours of direct sunlight per day, for vigorous growth and fruit production
        - They do not grow well in shady or cloudy conditions
- Impacts of Climate Change
    - Climate change poses significant threats to coconut cultivation, a vital agricultural sector in many tropical regions. Coconut palms are sensitive to environmental shifts, making them vulnerable to the effects of rising temperatures, altered rainfall patterns, and an increase in extreme weather events

- Prolonged heatwaves stress coconut trees, affecting their growth, flowering, and fruit production. High temperatures lead to faster maturity, potentially impacting kernel quality
- Disruptions in rainfall patterns, including both prolonged droughts and excessive rains damage coconut crops. Droughts lead to water scarcity, hindering growth and fruit production, while excessive rainfall cause root damage and drowning of palms. Example: Indonesia is facing challenges with both droughts and excessive rain affecting coconut harvests
- Coastal coconut plantations are vulnerable to rising sea levels, which lead to soil salinity. Saline soil reduces the ability of trees to absorb nutrients, impacting yields, and potentially causing trees to fall due to erosion
- Tropical cyclones and other extreme weather events are becoming more frequent and severe, causing physical damage to coconut trees and reducing productivity. Example: Typhoons have devastated plantations in the Philippines, one of the world's largest coconut producers
- Changing climatic conditions create favorable environments for the spread of pests and diseases, threatening coconut crops and yields. Some pests, like the scale insect *Aspidiotus destructor rigidus*, show increased reproductive efficiency, leading to rapid multiplication and extensive damage
- Economic and social implications
    - The climate-induced challenges have significant economic and social consequences. Reduced yields and damaged crops lead to inconsistencies in the coconut supply chain, impacting industries that rely on coconut derivatives and potentially leading to increased prices. For farmers, especially in developing countries, coconut farming is often a way of life, and crop losses can lead to long recovery periods and economic instability

- Mitigation and Adaptation strategies to Climate Change
  - To address these challenges and build resilience in the coconut industry, a variety of mitigation and adaptation strategies are being explored
    - Developing and utilizing drought-tolerant coconut varieties, implementing mulching to retain soil moisture, and practicing rainwater harvesting are crucial for managing water scarcity
    - Techniques like drip irrigation deliver water directly to the roots, minimizing waste
    - Employing organic farming practices, including the use of compost and biofertilizers help improve soil fertility and health
    - Planting mangroves near coastal areas can act as natural barriers against rising sea levels and soil salinity, while developing salt-resistant coconut varieties can further mitigate the effects of saline soils
    - IPM approach which combines methods like biological control (using natural predators) and reduced levels of pesticides to control pests, with regular monitoring and targeted interventions to minimize crop losses
    - In the long term, some regions may become unsuitable for coconut farming, potentially requiring a shift to higher altitudes or new geographical areas with more favorable conditions

13.3.2 Coffee *Coffea arabica* (Arabica coffee) and *Coffea canephora* (Robusta coffee)

- Climatic Requirements
- Coffee crops thrive in specific tropical and subtropical climates characterized by consistent temperatures, adequate rainfall, and appropriate altitudes

- Temperature
    - Arabica: Optimal temperature range is 15°C to 24°C (60°F to 75°F). Temperatures above 30°C (86°F) can negatively impact growth and bean quality
    - Robusta: Can tolerate slightly warmer temperatures, generally between 24°C and 30°C (75°F to 86°F)
- Altitude
    - Arabica: Grows best at higher altitudes, from 600 to 2,200 meters (2,000 to 7,200 feet) above sea level. Higher altitudes slow down the maturation process, leading to more complex flavors
    - Robusta: Grows well at lower altitudes, typically from sea level to 600 meters (2,000 feet)
- Rainfall
    - An annual rainfall of 1500mm to 2000mm (60 to 80 inches) is required
    - A distinct dry season is also necessary for the flowering and fruiting stages of coffee plants
    - Too much or too little rainfall negatively impacts growth and bean quality
- Sunlight
    - Coffee plants need a balance of sunlight and shade. Shade trees can help regulate temperature and moisture

- Impacts of Climate change
    - Climate change is significantly impacting coffee production worldwide. Rising temperatures, altered rainfall patterns, and increased frequency of extreme weather events are reducing the areas suitable for coffee cultivation and negatively affecting crop yields and quality. This poses a threat to the livelihoods of millions of coffee farmers and could lead to higher coffee prices for consumers
    - Reduced Suitable Land
        - Rising temperatures are shrinking the areas where coffee can be grown, particularly in traditional coffee-producing regions

   - While some new areas may become suitable at higher altitudes, the overall trend is a decrease in the total area of suitable land
- Decreased Yields and Quality
   - Warmer temperatures negatively impact coffee plant metabolism, reducing photosynthetic efficiency, and affecting bean ripening
   - Inconsistent rainfall patterns, with some regions experiencing drought while others face increased rainfall, disrupt harvest cycles and negatively impact bean quality
   - Extreme weather events like droughts and floods cause significant crop damage, leading to reduced yields and increased prices
   - Climate change can also favor the spread of pests and diseases that affect coffee plants, further impacting yields
- Economic Impacts
   - Lower coffee yields and reduced quality can lead to decreased income for coffee farmers, potentially forcing some out of business
   - The rising cost of production due to climate change adaptation measures and increased labor costs can also impact farmer livelihoods
   - Consumers can expect to see higher prices for coffee because of reduced supply and increased production costs
- Adaptation and Mitigation Strategies to Climate Change
   - The coffee industry is exploring various adaptation strategies, including developing more heat-tolerant coffee varieties and exploring agroforestry practices like growing coffee under shade trees
   - Shifting coffee cultivation to higher altitudes and diversifying coffee production regions are also being considered

- Consumers can support sustainable coffee farming practices by choosing beans from roasters and suppliers that work directly with farmers and prioritize environmental sustainability

13.3.3 Tea (*Camellia sinensis*)

- Climatic Requirements
- Tea thrives in warm and humid climates
  - Temperature
    - The ideal temperature range for tea cultivation is generally between 18°C and 29°C
    - While tea plants adapt to warmer temperatures, excessive heat (above 32°C) negatively impact photosynthesis
    - Temperatures should not drop below freezing for extended periods, as this damages the plants
  - Rainfall
    - A minimum annual rainfall of 1500mm is required, with some areas needing up to 2500-3000mm for optimal yield
    - The rainfall should be well distributed throughout the year, with a monthly average of at least 100mm during the growing season
    - Drought significantly reduces yield and increases dormant buds, while excessive rainfall leads to waterlogging and root rot
  - Humidity
    - High humidity, ideally between 70% and 90%, is essential for tea cultivation
    - Relative humidity below 60% negatively impacts tea growth and quality
  - Sunlight
    - Tea plants require sufficient sunlight for photosynthesis, but excessive sunlight is detrimental

- Altitude
  - Tea is grown at various altitudes, from sea level to over 2,100 meters, but higher altitudes with misty conditions are often associated with complex flavors

- Impacts of Climate Change
  - Climate change is significantly impacting tea production globally, affecting both the quantity and quality of tea. Rising temperatures, altered rainfall patterns, and increased frequency of extreme weather events like droughts and floods are major challenges for tea farmers. These changes lead to reduced yields, shifts in growing seasons, and even changes in the taste of tea
  - Impact on Yield
    - Climate change can lead to reduced tea yields due to factors like increased evapotranspiration, soil erosion, and pest infestations
    - Erratic rainfall, including both droughts and floods, can negatively impact tea production. Example: Drought can cause soil desiccation, while excessive rainfall can lead to waterlogging and root rot
    - While moderate temperature increases can sometimes boost yield, excessive heat can reduce productivity
  - Impact on Quality
    - Changes in temperature and rainfall can alter the chemical composition of tea leaves affecting their taste, flavor, and aroma
    - Heat and drought stress can impact on the levels of beneficial compounds like catechins and flavonoids in tea
  - Shifts in suitable areas
    - Climate change can alter the suitability of certain areas for tea cultivation, potentially leading to a shift in growing regions

- Increased pests and diseases
  - Higher temperatures can allow certain pests and diseases that attack tea plants to survive longer and reproduce more rapidly, posing a threat to tea bushes
- Socio-economic Impacts
  - The tea industry is vital for many regions, and climate change impacts can threaten the livelihoods of tea farmers, workers, and traders
  - Climate change impacts, coupled with other factors like low tea prices and rising production costs, can create economic hardships for the tea sector

- Mitigation and Adaptation Strategies to Climate Change
  - Incorporating trees into tea farms can help mitigate the effects of climate change by providing shade, improving soil health, and reducing erosion
  - Research on developing tea varieties that are more tolerant to drought, heat, and other climate-related stresses is crucial
  - Implementing efficient irrigation systems and water conservation practices can help address water scarcity issues
  - Adopting sustainable farming methods, such as integrated pest management, can reduce the reliance on chemical pesticides and fertilizers, which can also contribute to climate change

13.4 Major Vegetable Crops of the World

13.4.1 Tomato (Solanum lycopersicum)

- Climatic Requirements
- Tomatoes thrive in warm and sunny environments
  - Temperature
    - Optimal: 70-82°F (21-27°C) during the day and 62-64°F (16-18°C) at night
    - Germination: 65-85°F (18-29°C)

        - Fruit Set: Temperatures above 85°F (29°C) hinder fruit set
        - Growth: Temperatures above 25°C (77°F) reduce yield, especially when combined with high humidity and strong winds
    - Sunlight
        - Tomatoes need a minimum of six hours of direct sunlight daily
        - Full sun is ideal, but tolerate some light shade
    - Humidity
        - High humidity leads to increased pest and disease problems and fruit rot
    - Rainfall
        - Consistent watering is important, but avoid overwatering to prevent root rot
    - Day Length
        - Tomato is generally day-neutral, meaning their growth is not significantly affected by day length
- Impacts of Climate Change
    - Climate change is significantly impacting tomato crops worldwide, leading to reduced yields, decreased fruit quality, and disruptions in production. Rising temperatures, altered precipitation patterns, and increased frequency of extreme weather events like heat waves and droughts are key drivers of these negative impacts
    - Tomato plants are particularly sensitive to temperature fluctuations. High temperatures, especially above 29°C during the day and 21°C at night, can cause flower drop, reduced fruit set, and lower overall yields
    - High temperatures can also lead to smaller fruit size, uneven ripening, and softer fruit affecting both market value and processing quality

- Climate change can create conditions more favorable for certain diseases and pests that affect tomatoes. Example: Increased rainfall and humidity can promote the spread of bacterial pathogens, while warmer temperatures can lead to outbreaks of pests like spider mites
- The impacts of climate change on tomato production vary by region. Example: Some cooler regions may initially see some benefits from warmer temperatures, but ultimately, most tomato-growing regions will experience negative effects
- Drought and reduced water availability, which are becoming more common due to climate change, can further stress tomato plants and reduce yields
- The decline in tomato production can have significant economic consequences for farmers and the tomato processing industry, as well as potentially affecting global food security
- Examples
    - A study predicted a 6% decline in tomato yields in major growing regions like Italy and California by 2050
    - California, a major tomato producer, has already experienced lower yields in recent years due to high temperatures and drought
    - In Italy, research suggests that global warming will significantly impact tomato production, with some regions experiencing adverse weather conditions leading to stressed plants and poor harvests

- Mitigation and Adaptation Strategies to Climate Change
    - Greenhouse growers are adopting technologies like fog systems and improved climate control to mitigate the effects of high temperatures
    - Improving water-use efficiency through techniques like drip irrigation and using drought-tolerant tomato varieties can help address water scarcity

- o Developing and using climate-resistant tomato varieties is crucial for long-term sustainability
- o Optimizing planting and harvesting times can help avoid extreme weather events and minimize yield losses
- o Utilizing satellite data to monitor tomato crops and assess their health can help farmers make informed decisions and optimize resource management

13.4.2 Brinjal (*Solanum melongena*)

- Climatic Requirements
- Brinjal, also known as eggplant or aubergine, is a warm-season crop requiring specific climatic conditions to thrive and produce well
  - o Temperature
    - ▪ Optimal daytime temperature range: 20-27°C (68-81°F)
    - ▪ Ideal for fastest growth: 20°C - 30°C (68°F - 85°F)
    - ▪ Growth is significantly affected when temperatures fall below 17°C (63°F)
    - ▪ Highly susceptible to frost; temperatures below 13°C (55°F) harm the plants, and frost will damage the plants
    - ▪ Fruit setting is affected when temperatures dip below 15°C (60°F) or rise above 35°C (95°F)
    - ▪ Can be grown at elevations up to 1200 meters above sea level
  - o Sunlight
    - ▪ Requires full sun exposure, i.e., at least 6-8 hours of direct sunlight per day
    - ▪ Insufficient sunlight leads to yellowing of leaves and stunted growth
  - o Humidity
    - ▪ High humidity is preferred for storage

- Rainfall
  - Needs consistent moisture for optimal growth and fruit development
  - Requires at least an inch of water per week, with 2 inches being preferable, depending on factors like air temperature, wind, soil type, and the use of mulch
  - During hot, dry periods or low relative humidity, weekly irrigation may need to exceed 40 mm
  - In plains, irrigation should be every 3-4 days in hot weather and every 7-12 days in winter
  - Maintaining consistent moisture, especially during fruit set and development, is crucial to prevent oddly shaped fruits
  - Plastic mulch and avoiding overhead irrigation help manage water and prevent certain diseases
  - Drip irrigation is recommended for efficient water delivery
- Wind
  - Brinjal is sensitive to strong winds, so providing protection like windbreaks or mulching is important

- Impacts of Climate change
  - Global climate change is significantly impacting brinjal (eggplant) production, primarily through increased temperatures, altered rainfall patterns, and more frequent extreme weather events. These changes can lead to reduced yields, altered fruit quality, and increased susceptibility to pests and diseases
  - Brinjal thrives in warm conditions, but extreme heat can negatively affect fruit development and quality, leading to sunburned or scalded fruits. Higher temperatures can also shorten the plant's life cycle, reducing overall yield
  - Climate change can cause shifts in rainfall patterns, leading to both droughts and floods, which directly impact brinjal growth. Water stress, whether from drought or excessive moisture, can hinder photosynthesis, reduce nutrient uptake, and affect fruit quality

- Heat waves, floods, and droughts, becoming more frequent due to climate change, can cause significant damage to brinjal crop. These events can disrupt pollination, damage plants, and reduce overall productivity
- Changes in temperature and humidity can alter the distribution and behavior of pests and diseases, making brinjal more susceptible to infestations and infections
- Climate change can affect the nutritional composition of brinjal fruits, potentially leading to a decrease in important vitamins and minerals

- Mitigation and Adaptation Strategies to Climate Change
  - Developing brinjal varieties that are tolerant to heat, drought, and other climate-related stresses is crucial
  - Implementing efficient irrigation systems can help manage water stress and ensure adequate water supply during dry periods
  - Adopting integrated pest management strategies can help control pests and diseases while minimizing the use of harmful pesticides
  - Implementing climate-smart agricultural practices, such as conservation tillage and agroforestry, can improve soil health and water retention, making brinjal crops more resilient to climate change

13.4.3 Leafy Vegetables

- Climatic Requirements
- Leafy vegetables prefer cooler temperatures and consistent moisture, but specific requirements vary by type
  - Temperature
    - Cool-season crops: (lettuce, kale, spinach) prefer temperatures between 40°F and 70°F (4°C and 21°C)
    - Warm-season crops: (amaranth, malabar spinach) prefer temperatures above 65°F (18°C)

- Ideal temperatures: Lettuce and mustard greens thrive in 55-65°F (12-18°C), while bok choy prefers slightly warmer temperatures up to 75°F (23°C)
- Frost tolerance: Leafy greens tolerate light frosts (28°-32°F), but temperatures below 25°F are damaging
- Some leafy greens, like kale and collards, can tolerate warmer temperatures, while others, like mustard greens are more sensitive to heat and may bolt (begin to flower and set seed) in warm weather

- Sunlight
  - While leafy greens tolerate shade better than some other vegetables, they still need at least 5-7 hours of direct sunlight daily for optimal growth
- Moisture
  - Leafy greens require consistent moisture and are sensitive to drought
  - They need adequate water, especially during hot weather, to prevent bolting and maintain succulent growth
  - Supplemental irrigation may be necessary if rainfall is insufficient

- Impacts of Climate Change
  - Global climate change is significantly impacting leafy vegetable production and nutritional value. Higher temperatures, increased carbon dioxide levels, and altered precipitation patterns can reduce yields, affect nutrient content, and increase susceptibility to diseases. These changes can also affect the taste, texture, and marketability of leafy vegetables
  - Reduced Yields
    - High temperatures can lead to reduced growth rates and smaller plant sizes
    - Extreme weather events like droughts and heavy rainfall can severely impact leafy vegetable production

    - In some cases, higher carbon dioxide levels can initially increase yield, but this effect is often negated by the impacts of heat stress
- Nutritional Changes
    - Elevated carbon dioxide levels can decrease the concentration of essential minerals like calcium and antioxidant compounds in leafy greens
    - Increased temperatures can also affect the sugar and starch content of leafy vegetables, potentially impacting their nutritional value and taste
    - Higher carbon dioxide levels may also lead to reduced protein and nutrient levels in crops
- Physiological Disorders
    - High temperatures can cause sunburn and tipburn in leafy vegetables
    - Extreme heat can also lead to leaf scorching and water-soaked spots on leaves
- Increased Disease Susceptibility
    - Heat stress can weaken plants, making them more vulnerable to diseases and pests
    - Changes in temperature and humidity can create favorable conditions for the development of certain diseases
- Impact on Marketability
    - Visible damage like bruising, sunburn, or decay can reduce the market value of leafy vegetables
    - Disorders caused by heat stress or disease can lead to losses for farmers
- Changes in Taste and Texture
    - High temperatures can make leafy vegetables less tasty and crisp
    - Excess water from heavy rainfall can cause cracking and alter texture

- Mitigation and Adaptation Strategies to Climate Change
  - Researchers are working on breeding vegetable crops that are more tolerant to heat, drought, and other climate-related stresses
  - Using technologies like shade cloths, plastic mulches, and microbial inoculants can help mitigate the negative effects of climate change on leafy vegetable production
  - Implementing sustainable farming methods can help reduce greenhouse gas emissions and improve soil health, contributing to climate change mitigation and adaptation
  - Implementing efficient irrigation techniques and managing water resources effectively is crucial for mitigating the impacts of drought
  - Growing a variety of leafy vegetables can help reduce the risk of crop failure due to climate-related stresses

13.5 Impact of Climate Change on Pests and Diseases of Horticultural Crops

- Climate change is significantly altering the dynamics of pests and diseases in horticultural crops globally, leading to shifts in their geographic distribution, life cycles, and virulence. This poses a major threat to food security and the livelihoods of farmers
- Mango
  - Pests: Mango sudden decline and mango malformation are emerging issues linked to climate change. Studies suggest a link between climate change and mango pests and diseases, particularly influenced by regional factors and education levels of farmers. High temperatures can increase the feeding activity of pests like leafhoppers, but development can decline above certain temperature thresholds
  - Diseases: The warm and humid climate favors the development and spread of diseases like anthracnose, black mildew, and leaf blight in mango. Increasing rainfall and subsequent splashing of raindrops can facilitate the spread of microorganisms and increase pest and disease incidence

- Oranges
    - Pests: Warmer temperatures can accelerate the development and reproduction of pests like the navel orangeworm, potentially leading to more generations per year and increased crop damage. The invasive South American leaf miner may also overwinter more frequently outdoors due to warmer winter conditions
    - Diseases: Fungal infections like anthracnose and citrus canker thrive in humid conditions which can be exacerbated by climate change
- Grapes
    - Adverse conditions due to climate change intensify the likelihood of grape pests and diseases like mildew. Factors like warmer temperatures affecting insect overwintering and excessive heat and water scarcity contribute to these issues
- Coconut
    - Pests: Increased frequency and intensity of typhoons due to climate change create ideal breeding grounds for coconut rhinoceros beetles, an invasive species, leading to their proliferation and spread
    - Diseases: Changes in temperature and precipitation patterns can lead to water stress in coconut trees, making them more susceptible to disease outbreaks
- Coffee
    - Climate change can increase coffee leaf rust and pests like the coffee berry borer due to warmer temperatures, altered precipitation, and extreme weather. Changes in rainfall can favor root-knot nematodes. Bacterial blight is also a threat
- Tea
    - Rising temperatures, changes in rainfall, and increased soil erosion negatively impact tea yield and quality. Climate change is linked to new pests and diseases attacking tea bushes, and some fungal pathogens may expand their ranges

- Tomato
  - o Drought stress makes tomato plants more susceptible to pests like two-spotted spider mites, and warmer temperatures accelerate insect growth and reproduction. Diseases like late blight, leaf curl, and black spot can be exacerbated by climate changes, with high humidity and rainfall favoring fungal diseases
- Vegetables
  - o Climate change can promote pests and disease spread in vegetables, with drought stress increasing vulnerability to insect attacks. Elevated carbon dioxide, higher temperatures, and increased rainfall can worsen certain vegetable diseases and favor soilborne pathogens
- Other horticultural crops
  - o Fusarium wilt, a devastating disease of bananas, has spread globally due to trade and climate change
  - o Outbreaks of sigatoka disease in bananas due to favorable conditions like wetness duration is reported
  - o Increased prevalence of diseases like powdery mildew and scab in rainy seasons due to increased humidity and temperature
  - o Spread of fruit flies in heavy rainfall regions
  - o Impacts on fruit maturity and quality, with changes in the timing and severity of physiological disorders

13.5.1 Overall Impacts and Mitigation and Adaptation Strategies

- Climate change weakens host plants, making them more vulnerable to pests and diseases. Addressing these challenges requires climate-resilient crop varieties, integrated pest management (IPM), and adapting agricultural practices. Adaptation measures can include optimized irrigation, using tolerant cultivars, protective measures, planting trap crops or using biological control agents, improving soil management, and utilizing agroforestry

13.6 Addressing Climate Change Challenges of Horticultural Crops

- To address the climate change challenges individuals, families, communities, governments, organizations etc., have the best sciences available to plan for and implement climate-smart and resilient practices. Two critically important approaches which are available to managing climate resilience are mitigation and adaptation. These strategies are inter-related and must be pursued simultaneously as both can potentially improve resilience to the changing climate.

Chapter 14

Climate Change and Livestock and Humans

"The Vedas emphasize Ahimsa (non-violence), toward all living beings, recognizing them as spiritual entities subject to karma and rebirth. Cows, in particular, hold immense significance, symbolizing wealth, motherhood, and righteousness, while other animals are also revered as vehicles of gods and expressions of divine nature. This perspective fosters a deep interconnectedness between humans and the natural world, urging compassion and respect for all life"

14.1 Introduction to Climate Change and Livestock

- Two-thirds of domestic livestock are supported on rangelands, although in some countries a significant share of animal fodder also comes from crop residue
- The elevated temperature in arid and semi-arid rangelands is causing a manifold increase in potential evapotranspiration, leading to severe water stress conditions
- Because of the current marginality of soil water and nutrient reserves, some ecosystems in semi-arid regions may be among the first to show the effects of climate change
- Climate change has potential to exacerbate the loss of biodiversity in semi- arid regions thus affecting production of fodder for livestock

- The impact of climate change on livestock is negative in the humid and sub-humid tropics, particularly in the latter
- Heat stress has a variety of detrimental effects with significant effects on milk production and reproduction in dairy cows and swine fertility
- Moreover, warming in the tropics during warm months affects livestock reproduction negatively (reduced animal weight, dairy production, and feed conversion efficiency)
- Impacts, however, may be minor for relatively intense livestock production systems
- Livestock in humid areas in Africa are one way prone to diseases such as those carried by the tsetse fly and the other way reductions in the prevalence of tsetse in some current areas of distribution
- Because of the increasing trend for meat consumption, there is a higher demand for livestock feed. Production of feed grain in Asia may be affected by climate change
- In addition, the Earth's climate is undergoing unprecedented changes with profound implications for all living systems
- Livestock, vital for global food and livelihood security, are particularly vulnerable to these changes. The estimated annual U.S. livestock industry loss caused by heat stress alone falls between $1.7 and $2.4 billion
- India is a significant livestock producer. Studies show that during May and June, over 25 per cent of places experience a Temperature Humidity Index (THI) exceeding 85, indicating very high heat stress for livestock
- Therefore, there is a need to explore how these evolving climatic conditions exert a detrimental impact on the health and productivity of major livestock categories
- Climate change poses threat to the health and productivity of livestock and global food security

- The multifaceted impacts of the climatic factors on the health of economically important livestock species, including cattle, buffalo, sheep, goat, and poultry are occurring through direct and indirect mechanisms due to climate change
- The effects of climate change are enormous on feed intake and metabolism, reproductive efficiency, hormonal balance, behavioral adaptations, and immune system function, ultimately increasing susceptibility to diseases
- Understanding these intricate interactions is crucial for developing robust mitigation and adaptation strategies to safeguard livestock production systems and ensure sustainable animal agriculture in a changing climate

14.2 Climate and Livestock

- Bio-synoptics: The study of large-scale weather effects on biological processes, as shown by surface and upper air meteorological charts. Example: An influx of subtropical warm air at high levels may lead to an increase in number of clinical illnesses
- Biometeorology: The science that investigates the effect of individual weather phenomena on living processes. Example: When a lightning strikes some wood, it not only damages the wood tree it hits but may bring about death of surrounding trees
- Bio-climatology: The science which investigates the connection between climate and life
- Animals of agricultural importance depend on the variable microclimate of the upper soil and lower atmosphere
- Animals are directly affected by all-weather elements

14.2.1 Weather Elements and their Direct effects on Livestock

- Temperature: Temperature has the most significant direct impact on livestock productivity. Each species has a thermoneutral zone (TNZ) – a range of ambient temperatures where energy expenditure for maintaining body temperature is minimal. For dairy animals, the TNZ is typically 16-25°C. Beyond this, animals experience heat stress (HS), activating physiological and behavioral compensatory mechanisms

- Humidity: High humidity exacerbates the effects of high temperatures by reducing the efficiency of evaporative cooling (sweating and panting), making heat dissipation more challenging. A combination of temperatures higher than 37.7°C with over 50 per cent humidity has detrimental effects on cows and hogs
- Rainfall: Changes in rainfall directly impact the availability and quality of feed and water intake of livestock. Droughts lead to feed shortages and water scarcity, while floods can cause direct mortality, spread waterborne diseases, and damage infrastructure
- Extreme Weather Events: Increased frequency and intensity of events like cyclones, blizzards, and severe storms directly threaten livestock, leading to injuries, displacement, and mortality

Table. 14.1 Ideal Climatic Requirements for Farm Animals.

Species	Ideal Temperature (°C)	Relative Humidity (%)	Photoperiod (hrs)	Wind Flow / Ventilation Requirement
Cattle	15–25°C (optimum for reproduction)	60–70%	~12 hrs natural light	2–4 m/s; good air exchange in housing
Buffalo	13–18°C (cooler temps preferred)	60–75%	~10–12 hrs natural light	2.5–5 m/s; need for wallowing/sprinklers
Sheep	10–25°C (varies by breed)	50–60%	10–12 hrs (seasonal breeders)	2–3 m/s; dry, draft-free sheds in winter
Goat	15–30°C	50–60%	10–12 hrs (light-sensitive)	2–3 m/s; cross-ventilation in sheds
Pigs	16–22°C (sows), 18–25°C (boars, piglets)	60–70%	~12 hrs (not photoperiod-sensitive)	0.5–1 m/s in pens, avoid drafts on piglets

14.3 Impacts of Climate Change on Livestock

- Climate change impacts/affects livestock in four ways
  - The impact of changes on availability and price of fodder and feed grain
  - Impacts on livestock pastures and forage crops
  - The direct effects of weather and extreme events on animal health, growth, and reproduction
  - The changes in the distribution of livestock diseases
- In a world grappling with climate change, the role of livestock is multifaceted and involves both contributing to the problem and providing potential solutions
- Livestock production is a significant contributor to global greenhouse gas (GHG) emissions, particularly methane and nitrous oxide
- Ruminant animals like cattle and sheep naturally produce methane as a byproduct of digestion, releasing it primarily through burping
- Manure decomposition also releases methane and nitrous oxide, especially when stored under anaerobic conditions
- The production of feed crops for livestock, including the use of fertilizers and the associated land use changes, contributes to GHG emissions
- Expansion of pastures and feed crops for livestock can lead to deforestation and the release of stored carbon in forests and soils
- Livestock provides essential nutrients like protein, vitamin B12, iron, and zinc playing a crucial role in global food security
- Millions of people worldwide rely on livestock for their livelihoods especially in regions with marginal conditions for crop production
- Livestock also provides valuable services beyond food such as manure for fertilizer

- Changes in precipitation patterns, increased droughts, and heat stress can impact the availability and quality of feed and water resources for livestock
- Rising temperatures can lead to heat stress in livestock, impacting their productivity (milk and meat production), reproduction, and overall health
- Climate change can alter the geographic distribution and incidence of livestock diseases, potentially increasing outbreaks and impacting animal health
- Livestock are vulnerable to the impacts of extreme weather events like floods and storms, which can cause significant losses and damage to infrastructure
- Enhancing livestock productivity through better breeding, health management, and feeding practices can reduce the number of resources needed per unit of product, thereby reducing GHG emissions
- Practices like rotational grazing and agroforestry can improve soil health, sequester carbon, and enhance the resilience of grasslands to climate change impacts
- Implementing practices like composting, anaerobic digestion, or manure drying can reduce methane and nitrous oxide emissions from manure
- Developing and promoting livestock breeds that are better adapted to heat, drought, and disease can enhance the resilience of the sector to climate change
- Supportive policies and economic incentives are crucial for encouraging the adoption of climate-smart livestock practices, particularly in developing countries
- The role of livestock in the face of climate change is a complex and evolving issue
- Balancing the need for food security and livelihoods with the imperative to mitigate climate change requires a holistic approach that focuses on sustainable practices, innovation, and supportive policies

14.4 Heat Production and Heat Exchange in Livestock

- The loss of heat production must equal animal heat production. Otherwise, body temperature will rise or fall. This change declines the desired productive responses
- A resting (non-productive) dairy cow may produce approximately 400 kcal per hour when not lactating and about double this amount when lactating
- A simple equation or energy budget summarizes the balance between the energy produced by the animal and net loss of energy to the environment

$$H_p - H_s - H_w = H_r + H_c + H_e$$

Where,

- o H_p = production of heat by metabolism
- o H_s = storage of heat
- o H_w = Loss of heat when water and food are taken
- o H_r = Exchange of heat by radiation
- o H_c = Exchange of heat by convection and conduction
- o H_e = Loss of heat by evaporation of moisture

- Mostly, the direct effect of environment/weather upon animals is through this balancing of heat and the struggle to maintain a comfortable normal body temperature
- Farmers seek animals that tolerate heat and produce "Much" both at high and low temperatures
- Production by most of the animals declines markedly when temperatures exceed for a long time
- The critical temperatures range is 75° to 90 °F
- This critical temperature is modified by radiation, ventilation, and humidity

- Recently domestic animals have increasingly been confined to houses because of a possible greater efficiency of feed utilization and automation producers in herd management
- The confinement and sheltering of animals increase by the possibilities for excessive heating and the opportunities to remedy through engineering (shelter structure)
- All weather elements including diurnal cycle, length of day, acclimatization etc., individually and in combination must be considered in terms of animal responses

14.5 Effects of Heat Stress on Livestock

14.5.1 Effects of Heat Stress on Feed Intake and Digestion

- One of the most immediate and significant responses to heat stress is a reduction in feed intake, primarily dry matter intake. This is a survival strategy, as digestion and metabolism generate heat (heat increment of feeding)
- Mechanism: Elevated environmental temperatures negatively affect the appetite center in the hypothalamus. In lactating cows, feed intake can begin to decline at 25-26°C and drop by as much as 40% at 40°C. Dairy goats may see a 22-35% reduction, and buffalo heifers 8-10 per cent. Poultry also exhibits reduced feed intake, with a 9.5% drop observed in birds when ambient temperature increases from 21.1 to 32.2°C
- Rumen Physiology (ruminants): Heat stress alters the basic physiological mechanisms of the rumen. It can lead to
  - Reduced rumination and gut motility: Less chewing and slower passage rates
  - Changes in rumen microbial population and pH: A shift towards less fibrolytic bacteria, increasing the risk of subclinical and acute rumen acidosis due to decreased saliva production and lactic acid accumulation
  - Altered volatile fatty acid production: Decreased acetate production and increased propionate and butyrate, impacting energy utilization

- Negative Energy Balance (NEB): The combined effect of reduced feed intake and increased energy expenditure for thermoregulation leads to a state of NEB, resulting in body weight loss, decreased body condition score, and reduced productivity (milk yield, growth rate)

Table 14.2 Impact of heat stress on feed intake across livestock species.

Livestock Species	Temperature Range Causing Decline	Estimated % Decline in Feed Intake	Key Physiological Responses
Lactating Dairy Cows	>25-26°C (severe at >30°C)	Up to 40% (at 40°C)	Reduced appetite, altered rumen function, NEB
Dairy Goats	>10°C above TNZ	22-35%	Reduced appetite, increased maintenance requirements
Buffalo Heifers	Elevated temperatures	8-10%	Reduced appetite, impaired digestion

14.5.2 Effects of Heat Stress on Reproduction System of Livestock

- Climate change significantly compromises reproductive efficiency in livestock, leading to substantial economic losses
  - Females
    - Disruption of Estrus Cycles: Heat stress disrupts hormonal balance, affecting the pulsatile release of Gonadotropin-Releasing Hormone (GnRH) and subsequent Luteinizing Hormone (LH) and Follicle-Stimulating Hormone (FSH). This impairs follicular development, leading to anovulation or delayed ovulation, and poor expression of estrus due to reduced estradiol secretion
    - Reduced Conception Rates: Elevated body temperatures negatively impact oocyte quality, embryo development, and uterine environment, drastically lowering conception and pregnancy rates. Conception rates in dairy cows can drop by 20-27% during summer

        - Increased Embryonic Loss: Early embryonic death is more prevalent under heat stress conditions
        - Shorter Gestation Periods and Lower Birth Weights: Heat-stressed late gestation cows can have shorter gestation periods and produce calves with lower birth weights and impaired immune function
    - Males
        - Compromised Spermatogenesis: High scrotal temperatures negatively affect spermatogenesis, leading to reduced sperm concentration, motility, and increased morphological abnormalities
        - Reduced Libido: Heat stress can also decrease the sexual drive of male animals
        - Hormonal Changes: Testosterone levels can be lower during early exposure to hyperthermia

14.5.3 Effects of Heat Stress on Hormonal Changes in Livestock

- Heat stress elicits a cascade of endocrine responses aimed at maintaining homeostasis, but these often come at the cost of productivity
- Stress Hormones: Activation of the Hypothalamic-Pituitary-Adrenal (HPA) axis leads to increased secretion of glucocorticoids (cortisol). While cortisol helps the animal cope with stress, chronic elevation can suppress the immune system and negatively impact metabolism and reproduction
- Thyroid Hormones: Heat stress can lead to hypofunction of the thyroid gland, resulting in decreased levels of thyroid hormones (Triiodothyronine - T3 and Thyroxine - T4). These hormones are crucial for metabolic rate regulation, and their reduction is a mechanism to decrease metabolic heat production
- Reproductive Hormones: As mentioned, key reproductive hormones like GnRH, LH, FSH, estrogen, and progesterone are significantly altered, leading to the reproductive dysfunctions observed

- Growth Hormone: Corticotropin-Releasing Hormone (CRH) stimulated by heat stress can promote somatostatin, potentially leading to reduced growth hormone levels, impacting growth and milk production

14.5.4 Effects of Heat Stress on Behavioral Changes of Livestock

- Livestock exhibits various behavioral adaptations to mitigate the effects of heat stress, which can also impact their productivity and welfare
- Seeking Shade/Cooling: Animals will actively seek shade, whether natural (trees) or artificial (sheds). They may also increase their use of wallows (buffalo), sprinklers, or misters if available
- Reduced Activity and Resting: Animals become less active during the hottest parts of the day, reducing their movement and grazing time to minimize heat production. This can lead to decreased foraging efficiency
- Increased Water Intake: To compensate for fluid loss through sweating and panting, animals increase their water consumption
- Altered Feeding Patterns: Animals may shift their feeding to cooler times of the day (early morning or late evening) to avoid the heat increment of feeding during peak temperatures
- Aggression/Lethargy: Severe heat stress can lead to lethargy and, in some cases, increased aggression or changes in social hierarchy
- Panting and Open-Mouth Breathing: These are clear indicators of heat stress, aimed at increasing evaporative cooling

14.5.5 Effects of Heat Stress on Immune System and Disease Susceptibility of Livestock

- Heat and humidity impact animals' immune systems making it more susceptible to disease and stress
- Unfavorable temperatures (too cold/hot) lead to an increased heat production by the animal i.e., there is more loss of energy, and in consequence less energy remains for production at the same level of energy intake, and the efficiency of energy utilization deteriorates

- Climate change compromises the immune system of livestock, making them more vulnerable to infectious and non-infectious diseases
- Immune Suppression: Chronic stress, particularly heat stress, leads to elevated cortisol levels which are known as immunosuppressants. This reduces the efficacy of the immune response, making animals more susceptible to various pathogens
- Increased Incidence of Existing Diseases
    - Mastitis: Increases in dairy animals during hot and humid weather, often associated with increased fly and tick populations
    - Foot and Mouth Disease (FMD) and Hemorrhagic Septicemia (HS): The frequency of these diseases is likely to be higher due to climate change
    - Respiratory Diseases: Cattle exposed to chronic heat stress show higher rates of respiratory diseases
- Emergence and Spread of Vector-Borne Diseases
    - Vector Proliferation: Increased temperatures and altered precipitation patterns can create more favorable conditions for the breeding, development, and survival of disease vectors (ticks, mosquitoes, flies)
    - Expanded Geographical Range: Vectors and the pathogens they carry can expand their geographical distribution into previously unaffected areas as climatic zones shift. Tick-borne diseases like Lyme disease (sheep) and Anaplasmosis (cattle) are a concern
    - Accelerated Pathogen Development: Higher temperatures can accelerate the development rate of certain pathogens or parasites that have external life cycle stages
    - Waterborne Diseases: Increased rainfall and flooding can lead to higher incidences of waterborne diseases like Leptospirosis. Drought conditions can also concentrate animals around limited water sources, increasing transmission rates of diseases like Q fever

14.6 Way Forward

- The evidence overwhelmingly indicates that climate change is already exerting a profound and detrimental impact on the health and productivity of livestock worldwide. From directly inducing heat stress and altering metabolic functions to indirectly facilitating the spread of diseases and compromising reproductive success, the challenges are multifaceted and interconnected. The recent data underscores the urgency of addressing these issues
- For the international community, understanding these mechanisms is the first step towards developing effective mitigation and adaptation strategies. Future research and interventions must focus on
  - Developing Climate-Resilient Breeds: Genetic selection for heat-tolerant, disease-resistant, and climate-adaptive livestock. Tropical breeds like Bos indicus cattle show greater resistance to heat stress
  - Improved Housing and Management Practices: Implementing technologies such as advanced ventilation, cooling systems (fans, sprinklers), and adequate shade provision
  - Nutritional Interventions: Adjusting diets to provide higher quality food during stress periods, supplementing with vitamins, minerals, and antioxidants, and optimizing feeding times
  - Enhanced Disease Surveillance and Control: Strengthening early warning systems for climate-sensitive diseases and developing effective vaccination and treatment protocols
  - Sustainable Water Management: Ensuring access to clean and sufficient water resources, especially during periods of drought
  - Policy and Economic Support: Implementing policies that support farmers in adopting climate-smart livestock practices and providing financial incentives for adaptation measures

- Integrated Approaches: Recognizing the complex interplay, solutions must integrate veterinary science, animal nutrition, genetics, agrometeorology, and socioeconomic considerations

- By proactively addressing the impacts of climate change on livestock health, more resilient and sustainable animal agriculture systems, ensuring food security and the well-being of both animals and humans in a changing world can be achieved

14.7 Impacts of Climate Change on Fisheries

- The impact of climate change on global fisheries includes major disruptions to marine ecosystems, leading to shifts in fish populations, reduced productivity, and significant socioeconomic challenges for coastal communities
- Ecological impacts
    - Warming oceans: Rising sea temperatures cause many fish species to shift their distribution toward cooler, higher latitudes or deeper waters. This creates mismatches in the food web as predators and prey move at different rates, disrupting traditional ecosystems
    - Ocean acidification: Increased absorption of atmospheric carbon dioxide is making the ocean more acidic. This is particularly harmful for calcifying organisms like coral, oysters, and clams, as it hinders their ability to build shells and skeletons. The destruction of coral reefs, which shelter 25% of all marine life, threatens biodiversity and the food supply for millions
    - Habitat loss: Extreme weather, sea-level rise, and warming waters destroy vital coastal habitats, such as wetlands and coral reefs, that serve as nursery grounds for many commercially important species
- Socioeconomic impacts
    - Shifting catch patterns: Fish migrating to new areas disrupt established fisheries, forcing fishermen to travel farther,

target new species, or risk economic losses. In some cases, fishers in higher latitudes may benefit from new species, while those in warmer regions face declines

- Reduced stock productivity: Climate change can reduce fish growth rates, alter their reproductive cycles, and lower overall stock abundance, leading to smaller, and less productive populations. This decreases the number of fish available for catch and consumption
- Increased vulnerability for communities: Coastal and island nations that are highly dependent on fisheries for food and income are among the most vulnerable. Indigenous communities that rely on traditional fishing practices face increased risks due to changing conditions
- Conflicts over resources: As fish stocks cross geopolitical boundaries, disputes on fishing rights and quotas can arise between countries, making international fisheries management more complex

- Mitigation and Adaptation Strategies
    - Improve fuel efficiency: Use more efficient diesel engines, properly maintain propellers and hulls, and optimize fishing routes to reduce fuel use
    - Reduce fishing effort: Manage stock sustainably to rebuild depleted populations, which makes fishing more efficient and reduces fuel needed per catch. This involves reforming harmful subsidies that encourage overfishing
    - Shift fishing methods: Encourage fishing techniques with lower emissions, such as moving away from fuel-intensive bottom trawling
    - Invest in aquaculture: Expand sustainable marine aquaculture, which can have a smaller carbon footprint than some wild-capture fishing or land-based protein sources
    - Implement adaptive management: Develop flexible, science-based management plans that can quickly respond to shifting stock distribution, altered productivity, and ecosystem changes

- Adopt climate-informed spatial planning: Use oceanographic data and predictive models to understand and plan for shifts in species habitats. This includes designating and enforcing Marine Protected Areas (MPAs) and temporary fishing closures
- Diversify livelihoods: Support fishing communities in diversifying their income to reduce dependence on a single species. This can involve adopting new fishing technologies
- Prioritize equity: Empower vulnerable communities, including small-scale fishers, in adaptation planning to ensure fair access rights and livelihood security
- Invest in innovation and technology: Use data and technology, including electronic monitoring and early warning systems for harmful algal blooms, to help fisheries make more informed and timely decisions

14.8 Impacts of Climate Change on Poultry

- Climate change harms global poultry through direct heat stress on birds and indirect impacts on resources and disease. This reduces production, increases costs, and threatens food security, especially in tropical and arid regions
- Direct impacts on poultry
    - Heat stress: Rising temperatures and humidity overwhelm a bird's natural cooling, causing physiological stress. The effects are particularly severe in fast-growing modern breeds
    - Reduced productivity: Heat stress decreases feed intake and nutrient absorption, leading to lower growth rates, smaller body weight, and poor feed conversion ratios in broilers
    - Lowered egg quality and production: Laying hens produce fewer eggs, and the ones they do lay are smaller and have thinner shells
    - Impaired reproduction: Extreme temperatures negatively affect fertility and egg hatchability in breeding stock

- Increased mortality and disease: Heat stress weakens birds' immune systems, making them more susceptible to infectious diseases. It also directly causes higher mortality rates

- Indirect impacts on the industry
  - Feed scarcity and cost: Climate-related events like droughts, floods, and excessive heat reduce crop yields for key feed ingredients such as corn and soybeans. This drives up feed costs and can lead to malnutrition in flocks
  - Water shortages: Higher temperatures increase water demand, while changing precipitation patterns can cause scarcity or flooding, which reduces water quality
  - Infrastructure damage: Extreme weather events, such as hurricanes and floods, can destroy poultry housing, disrupt transportation, and cause power outages, crippling the supply chain
  - Higher operating expenses: Rising temperatures require increased energy use for ventilation and cooling systems, boosting electricity costs for farmers
  - Increased disease risk: Climate change can alter the migratory paths of wild birds, potentially introducing new diseases to domestic poultry populations
- Potential solutions, mitigation and adaptation Strategies
  - Developing heat-tolerant poultry breeds through genetic selection
  - Improving housing with better ventilation, cooling systems, and insulation
  - Adjusting feeding practices, such as feeding at cooler times of the day
  - Using nutritional additives to help birds cope with heat stress

- Improve feed efficiency: Since feed production is a major source of GHG emissions, optimizing feed formulas reduces the amount needed per unit of meat or eggs. Using alternative ingredients like insect meal or algae also lowers the environmental impact
- Sustainable manure management: Proper handling of poultry litter reduces methane and nitrous oxide emissions. Strategies include composting, using it as fertilizer, or turning it into bioenergy via anaerobic digestion
- Increase energy efficiency: On-farm energy consumption for heating, ventilation, and lighting can be reduced by insulating houses, installing energy-efficient systems, and transitioning to renewable sources like solar or wind power
- Upgrade housing systems: Improving insulation, ventilation, and cooling technologies (fogging, misting systems) protects flocks from heat stress. Using natural ventilation systems and reflective roofing also helps regulate house temperature
- Adjust management practices: Modifying stocking densities, providing shade, and altering feeding schedules to cooler parts of the day can help birds cope with high temperatures
- Employing genetic and nutritional strategies
    - Genetics: Breeding for heat-tolerant and disease-resistant traits enhances resilience. This includes using genes from indigenous, locally adapted breeds and research on thermotolerance markers
    - Nutrition: Adding antioxidants, probiotics, and specific amino acids to feed helps birds handle stress. Providing constant access to clean and cool water is also critical

- Build resilient infrastructure: Investing in durable, flood-resistant housing and better biosecurity measures minimize damage and disease risk from extreme weather events
- Enhance resource management: Implementing water conservation techniques like rainwater harvesting addresses potential water scarcity issues. Diversifying feed sources reduces reliance on climate-sensitive crops

14.9 Effects of Climate Change on Humans

14.9.1 Effects of Weather Elements on Humans

- Rainfall: Prolonged periods of heavy rain increase the possibility of vector borne spread of diseases
- Temperature: Even small differences from seasonal average temperatures are associated with increased illness. Temperature extremes can also worsen chronic conditions including cardiovascular, respiratory and Cerebro-vascular diseases and diabetes related conditions. Heat waves cause heat creams, and heat stroke. Heat crams and heat exhaustion are experienced between 32 to 40 °C
- Pressure: The most reported result of changes in barometric pressure on human health is associated with headaches and migraines. As the pressure decreases, the amount of oxygen available to breathe also decreases
- High humidity: High humidity can contribute to feelings of low energy and lethargy; hyperthermia (over heating). The overheating is a result of human body's inability to effectively let out heat. Under high humid weather conditions sweat cannot evaporate leaving human bodies feeling hot and sticky. Consequently, to cool off the human body must work even harder. This results in excessive sweating, increased blood circulation and enhanced respiration. The humidity range most ideal for human beings is 30 – 50%
- ENSO: This is associated with droughts and floods. The effect of ENSO on cholera risk in Bangladesh and malaria epidemics in parts of South Asia and South America have been well established. Mosquito borne and rodent borne diseases can be triggered by extreme weather conditions

14.9.2 Humans and Microclimate

- The climate in completely or partially enclosed spaces, especially used by man and animals is known as "Crypto climate"
- The air breathing has a profound influence on human wellbeing. Electrical charges of air such as ions, spheric and electrical fields can affect human digestive and nervous systems
- In much the same way as plants and animals, man reacts unconsciously in his first response to favorable or unfavorable microclimatic conditions
- Every building constructed displaces the original climate of its site, creating a warm, sunny, and dry climate with a Southern exposure on the one hand, and a shady, cold and damp Northern climate on the other. History has taught more than once that increasing urbanization has brought great damage to the heat and water budgets of a land
- The first type of microclimate to be considered is that prevailing in man's means of transport, viz., cars, trains, and aircraft
- Men know how to reduce the adverse effects of climate on their buildings and how to create an appropriate climate in their dwellings
- In a comprehensive report it was mentioned that man can feel comfortable with a temperature between 18 and 32 °C
- The climate of a room depends on the situation of house and on the position of the room inside the house. The direction in which it faces, its height above the seal level, the arrangement of windows and the immediate neighbors decide the atmospheric conditions to which the dwelling is exposed

14.9.3 Effects of Climate Change on Human Health

- The impact of weather on human health is a well-known fact, yet, neglected in the past
- Bioclimatology, a vast field of medical knowledge, has only been developed in the past few years

- The weather has profound effect on human health
- The changes in birth rates, outbreaks of pneumonia, influenza, bronchitis, etc. in addition to morbidity effects linked to pollen concentrations and high pollution effects
- Also, health effects include increased respiratory and cardiovascular diseases; injuries and premature deaths related to extreme weather events
- Changes in prevalence and geographical distribution of food and water born illnesses and other infectious diseases, threats to mental health, etc.
- Greenhouse gases: The greenhouse gases contribute to respiratory diseases from smog and air pollution. Global warming will cause infectious diseases such as malaria, cholera, etc.
- Climate change harms farmers/humans through disruptions to the global food system
- Farmers face direct risks to their livelihoods from extreme weather, pests, and water scarcity, while the broader population experiences higher food prices, reduced nutritional quality, and increased food insecurity
    - Reduced yields: Rising temperatures, droughts, floods, and new pests cause significant crop and livestock losses
    - Extreme working conditions: Increasing heat waves lead to unsafe and grueling labor for farmers and farm workers
    - Loss of income: Crop failure and livestock deaths result in income losses, particularly for small-scale farmers in vulnerable regions
    - Increased food prices and insecurity: Declining agricultural productivity leads to higher food costs and greater hunger, especially for low-income and vulnerable communities
    - Reduced nutrition: Higher carbon dioxide levels can lower the protein, zinc, and iron content in staple crops like rice and wheat

- Health threats: Climate-related issues like water contamination, air pollution from wildfires, and higher pesticide use pose widespread health risks
- Mental health stress: The uncertainty of food supply and livelihoods contributes to anxiety and depression

14.10 Addressing Climate Change Challenges on Livestock and Humans

- The individuals, families, communities, governments, organizations, etc., have the best sciences available to plan for and implement climate-smart and resilient practices. Two critically important approaches which are available to managing climate resilience are mitigation and adaptation. These strategies are inter-related and must be pursued simultaneously as both can potentially improve resilience to the changing climate.

Chapter 15

Climate Change, Food Production and Food Security, and Role of Agrometeorology

"Food is divine, essential for life, and deeply connected to spiritual well-being. Vedic agriculture emphasizes sustainable practices, reverence for nature, and equitable food distribution. Sharing and conscious consumption, rooted in ethical principles like Ahimsa, ensure food security and foster harmony"

15.1 Introduction to Food Production and Food Security

- Food security is a critical global challenge, intertwined with various interconnected factors
- Addressing it is crucial for a healthy, stable, and prosperous future
- Food availability refers to the overall supply of food and a region's or nation's capacity to produce and distribute it
- Food availability can be affected by high import costs or poor transportation links

- Food access addresses whether individuals and households can physically and economically obtain enough food to maintain a healthy diet
- Access is influenced by income levels, food prices, transportation options, and social factors
- Poverty significantly limits food access and increases vulnerability to food price spikes
- Utilization focuses on the nutritional quality of the food consumed and the body's ability to absorb and utilize nutrients effectively
- Utilization also involves factors like safe food preparation and storage, access to clean water and sanitation, health conditions, and nutritional education
- Food safety is a crucial aspect of utilization, as contaminated food can lead to illness and hinder nutrient absorption
- Stability concerns the consistency of food availability, access, and utilization over time
- Stability can be threatened by climate change, natural disasters (droughts, floods), economic shocks, political instability, conflict, and disruptions to the food distribution system
- Sustainability emphasizes the long-term ability of food systems to provide food security without compromising the environmental, social, and economic basis for future generations
- Sustainable agriculture practices, water quality regulations, and climate change mitigation are essential for maintaining healthy and productive food systems
- Agency refers to the ability of individuals and groups to make their own choices about food production, consumption, and to influence food system policies. This includes aspects of food sovereignty and ensuring fair trading practices
- Food security is essential for ensuring access to enough safe, healthy, and nutritious food for everyone, which promotes good health and prevents malnutrition in all its forms
- Malnutrition can lead to stunted growth, cognitive impairment, increased susceptibility to disease, and reduced productivity

- A secure food system contributes to economic growth by supporting a productive workforce, reducing healthcare costs associated with food-borne illnesses, and fostering market stability by mitigating price volatility
- Food security, by fostering market stability and reducing hunger, can also contribute to the development of agricultural markets and economies, particularly in developing countries
- Food insecurity can contribute to social unrest, conflict, and instability
- Ensuring food security is a matter of social equity, guaranteeing access to nutritious food regardless of income, race, or gender, and addressing inequalities
- Sustainable food systems are vital for protecting the environment by preserving biodiversity, reducing greenhouse gas emissions, and conserving natural resources
- Sustainable agriculture is a crucial pillar of food security, contributing to maintaining environmental sustainability

15.2 Homes of Evolution of Agriculture for Food Production and Security in the "Old and New" World

- Researchers have identified seven (7) homes of evolution of agriculture in the "Old" world and three (3) in the "New" world
    - In the "Old" world
        - The oldest center of diversity and origin of cultivated species of plants in Southeast Asia is the areas comprising Thailand, Malaysia and adjoining areas for rice, sugarcane, banana, coconut, cardamoms, and turmeric
        - Second oldest center was the Fertile Crescent in West Asia for wheat, forages, olives, grapes, and almonds
        - Third was South India for rice, jute, *desi* cotton, pulses, mango, citrus species, and black pepper

        - Fourth was China for soybean, apricot, litchi, and some citrus species
        - Fifth was northern India, Afghanistan, and Iran for bread wheat, rye, linseed, and walnut
        - Sixth was the center bordering the Mediterranean for oats, temperate zone fruits, vegetables, cabbage, and cauliflower
        - Seventh was Ethiopia for wheat, sorghum, castor, barley, coffee, and watermelon
    - In the "New" world
        - Mexico for maize, sweet potato, cotton, chilies, guava, pumpkin, and beans
        - Peru and Bolivia for potato, tomato, and tobacco
        - Argentina and Brazil for rubber, cashew nut, cocoa, groundnut, and pineapple

15.3 Definitions

- Food security: A situation in which all people, at all times, have physical and economic access to sufficient, safe and nutritious food that meets their dietary needs and food preferences for an active and healthy life. This definition points to four dimensions of food security viz., availability of food, accessibility of food (economically and physically), utilization of food (the way it is used), and assimilated by the human body
- Moderate food insecurity: A level of severity of food insecurity at which people face uncertainties about their ability to obtain food and have been forced to reduce, at times during the year, the quality and/or quantity of food they consume due to lack of money or other resources. It refers to a lack of consistent access to food, which diminishes dietary quality and disrupts normal eating patterns. It is measured with the Food Insecurity Experience Scale (FIES)

- Severe food insecurity: A level of severity of food insecurity at which, at some time during the year, people have run out of food, experienced hunger and at the most extreme, gone without food for a day or more. It is also measured with the Food Insecurity Experience Scale

15.4 Sustainable Development Goals

- Sustainable Development Goals (SDGs): The 17 global goals for development for all countries established by the United Nations through a participatory process and elaborated in the 2030 Agenda for Sustainable Development, including ending poverty and hunger; ensuring health and well-being, education, gender equality, clean water and energy, and decent work; building and ensuring resilient and sustainable infrastructure in cities, consumption; reducing inequalities; protecting land and water ecosystems; promoting peace, justice and partnerships; and taking urgent action on climate change

15.4.1 Targets under the Sustainable Development Goals (SDG) of Climate Action for Food Production and Security

- Strengthen resilience and adaptive capacity to climate related hazards and natural disasters
- Integrate climate measures into national policies, strategies, and planning
- Build knowledge and capacity to meet climate change
- Implement the UNFCCC
- Promote the mechanisms to raise capacity for planning and management internationally

15.5 Significance of Climate Change on Food Production and Food Security

- Climate change is the principal source of fluctuations in global food production and food security
- It is also a major threat to adequate nutrition in the twenty-first century

- Climate change is influencing agroecosystems and the health of humans
- Climate change also has the potential to affect the selection of crop production and animal rearing locations, sowing and production of crops, livestock rearing, poultry and fishery systems etc., at local, national, regional, and global scales
- Ending hunger, achieving food security and improving nutrition are the priorities of the Sustainable Development Goals (SDGs)
- Climate change is a major challenge in this direction and making the task more difficult for food security, ending hunger, and malnutrition
- The effects of climate change on agroecosystems are already severe and widespread, and ensuring food security in the face of climate change is among the most daunting challenges facing humankind

15.6 Impacts of Climate Change on Food Production and Food Security in different Regions of the World

- Climate change has reduced food security and affected water scarcity, hindering efforts to meet sustainable development goals
- Although the overall agricultural productivity has increased, climate change has slowed the growth over the past 50 years globally, with related impacts mainly in mid and low latitude regions but positive impacts in high latitude regions
- Ocean warming and ocean acidification have adversely affected food production from fisheries and shellfish in some oceanic regions
- Roughly half of the world population currently experience severe water scarcity for at least part of the year due to a combination of both climatic and non-climatic drivers
- Population and food needs
    - o There are 836 million people living in extreme poverty

- Around 70 per cent of the very poor live in rural areas, most of them depend on agriculture for their livelihoods
- Approximately 500 million small holder farms in the developing world are supporting almost 2 billion people
- In Asia and sub- Saharan Africa these small farms produce about 80 per cent of the food consumed
- Between 660 and 820 million people (workers and their families) depend totally or partly on fisheries, aquaculture and related industries as a source of income and support
- In addition, rural poor often depend partly on forests for their livelihood

- The South Asia comprise eight countries viz., India, Afghanistan, Bangladesh, Bhutan, Maldives, Nepal, Pakistan and Sri Lanka
- This is a large land mass of 1.62 billion hectares, and the agroecosystems are strongly impacted by climate change
- Climate change has multi-dimensional impacts on agroecosystems including increases in temperature, decline in freshwater availability, sea level rise, increased frequency and intensity of extreme events, and shifting of cropping zones
- These impact agriculture and related food sector as well as general economies, societies, and the environment
- Small land holders (less than 2 ha.) and resource poor subsistence farmers predominate and contribute less to carbon dioxide emissions
- In many countries the per capita land area is less than 0.1 ha. and is rapidly decreasing because of conversion of land to non-agricultural uses, soil degradation and continued population growth
- The serious problems of soil degradation and desertification are exacerbated by climate change through accelerated erosion, fertility depletion, salinization, and acidification
- Subsistence agriculture, characterized by low productivity and extractive farming, is extremely vulnerable to climate change resulting in reduced food production
- Rains failed in central and eastern India in 1984, triggering one of the worst droughts

- The 1973 drought in Ethiopia claimed 2, 00,000 human lives
- In 1987-88 drought in Philippines was quite serious and 46 of the 78 provinces were proclaimed to be in calamity
- Widespread and severe vulnerability in Mexico to droughts, floods, and frosts in twentieth century are associated with climate change. More than 90 per cent of losses in Mexican agriculture are from droughts. In 1970, parts of northern Mexico lost more than 35 per cent of area planted due to drought
- In Niger, droughts occurred between 1910- 1920; 1940-1950; 1968-1973; and 1976-1984.
- Due to global warming the EL Nino, La Nina, and ENSO's are causing the most significant effects on climate change
    - Both the range and intensity of ENSO events have been much higher since 1950
    - Hot extremes have become more frequent and more intense
    - Oceans have been warming at rapid pace since the 1970s, and ocean acidification is intensifying
    - Effects on frozen areas of the planet
        - Global retreat of glaciers since the 1990s
        - 40% decrease in Arctic Sea ice since 1979
        - Decrease in spring snow cover since the 1950s

15.7 The FAO Report on the Cascading Effects of Climate Change on Food Security and Nutrition

- Some of the problems associated with climate change are emerging gradually
- Action is urgently needed to allow enough time to build resilience in agricultural production systems
- Plants, animals, and ecosystems generally adapt to the prevailing climatic conditions
- When these conditions change then the plants and animals will be impacted, some will become less productive, or even disappear

- There is thus a cascade of risks from climate changes to agroecosystems, to agricultural production, to economic and social consequences and finally to food security and nutrition
- Climate change affects food security in all its dimensions: access, availability, utilization and stability
- "Climate change" impacts " Agroecosystems" through changes in atmosperic composition, land degradation, ocean acidification, temperature, sea level rise, precipitation and water availabilty **(Fig.15.1)**

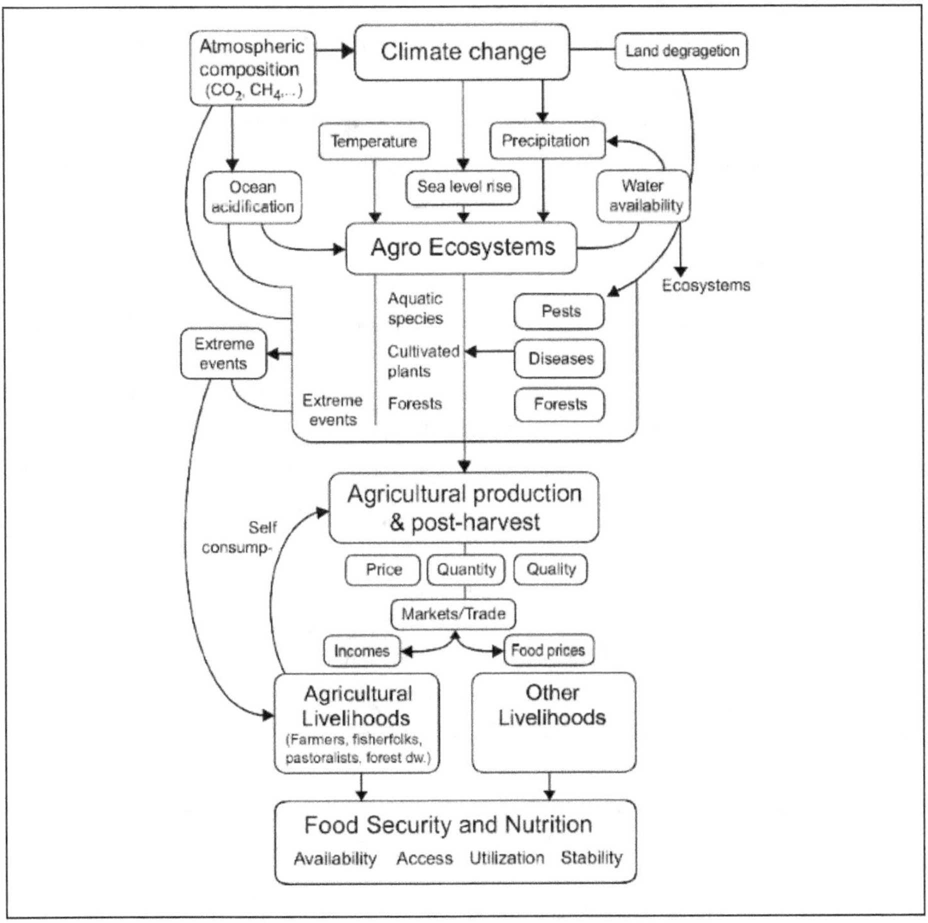

Fig. 15.1 Schematic representation of the cascading effects of climate change impacts on food security and nutrition.

Source: FAO (2015)

- As a consequence, the "Agroecosystems" are negatively afected by extreme events, occurance of pests and diseases on acquatic species, cultivated plants, livestock and forests
- In turn, the "Agroecosystems" impact " Agricultural production and post harvest activities" and results in fluctuations in prices, quantity, quality of agricultural products. This again impacs markets/trade
- The"Markets/trade" affects "Agricultural and other livelihoods" through incomes and food prices
- At the end, as stated above, "Food security and nutrition (availability, access, utilization, stability)" are threatned

15.8 The IPCC Report on Human Dimention of Climate Change at Nexus of Climate Change, Climate Hazards, and Socio Economic Development

- The impacts of climate change have significantly affected livelihoods and living conditions, especially the poorest and most vulnerable and will continue to undermine development during the coming century **(Fig:15.2)**

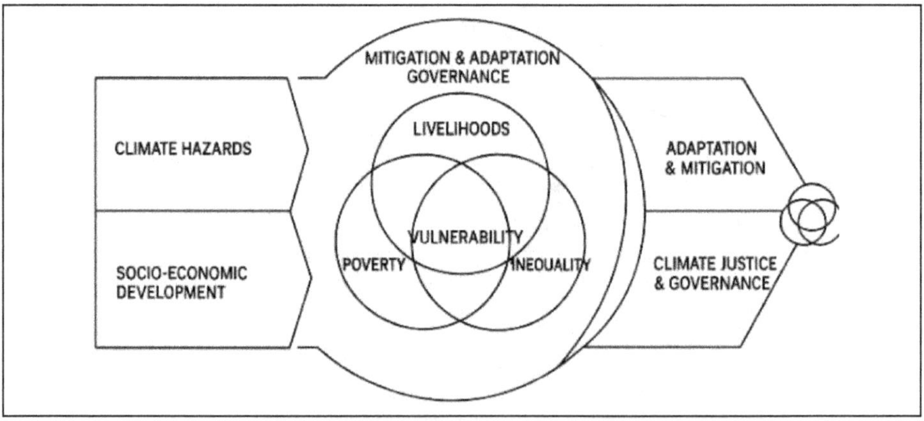

Fig. 15.2 Human dimension of climate change at nexus of climate change, Climate hazards and socio economic development.

Source: IPCC 2021,2022,and 2023

- The societal consequences of climate change and related hazards in terms of adverse and irreversible consequences for the most vulnerable were assessed
- To understand societal consequences of climate change, impacts through the perspective of vulnerability, poverty, and livelihoods of people were assessed
- It was identified why climate events trigger sudden and slow-onset disasters, and how the most severe, acute, and chronic impacts cause and deepen human suffering
- The issues of climate justice were also examined. Understanding and engaging in climate justice requires a plural focus on the historical social and institutional relations and inequalities that produce climate change, cause people to be vulnerable to climate hazards and shape responses to them
- An assessment of observed impacts on the poor and their strategies for adaptation carries important lessons for inclusive, broad-based solutions to climate change
- The linkages between climate change, specific climate-related hazards and impacts on multidimensional poverty, vulnerability and livelihoods were also examined
- The assessment reveals that successful adaptation is not solely a question of levels of funding, but depends on broader institutional design that determines societal development and enabling conditions for adaptation to and mitigation of climate change
- An assessment of enabling conditions for adaptation supports the finding that more convergent, integrated, and comprehensive approaches to adaptation are needed
- It was concluded that climate justice requires consideration of the legal, institutional and governance frameworks that significantly determine whether adaptation is successful in addressing the needs of the poor

15.9 Social and Economic Considerations in Climate Change Studies

- Human experience and simple prudence dictate greater efforts to incorporate climate considerations into the way natural resources can be managed in food production and distribution
- Research on how farmers perceive and react to climate change has generally been neglected
- Research work on characterizing agricultural environments in ecological and social terms needs to be reinforced so that the potential for sustainable management of agricultural systems to meet the changing demands for food in the face of climate change can be better understood
- The greatest need is to adapt technology to specific circumstances
- Adaption would be speeded up if farmers are seen as partners to be consulted earlier in the process
- Also, different types of agrometeorological information in this connection are useful in different areas
- Local communities must be involved in collection, analysis, and exchange of information to address their needs
- Location specific research by people sensitive to local views and traditions can improve scientific understanding of how farmers respond to climate change
- Follow up activities should explore how whole season and within season agrometeorological information can be effectively communicated to the farmers
- For this, agro meteorologists may collaborate with experts in mass communication and in village level field studies

15.10 Linkages among Food Production, Environmental Issues, and Climate Change

- The global food system faces a complex web of interconnected challenges spanning food production, environmental issues, and climate change **(Fig:15.3)**

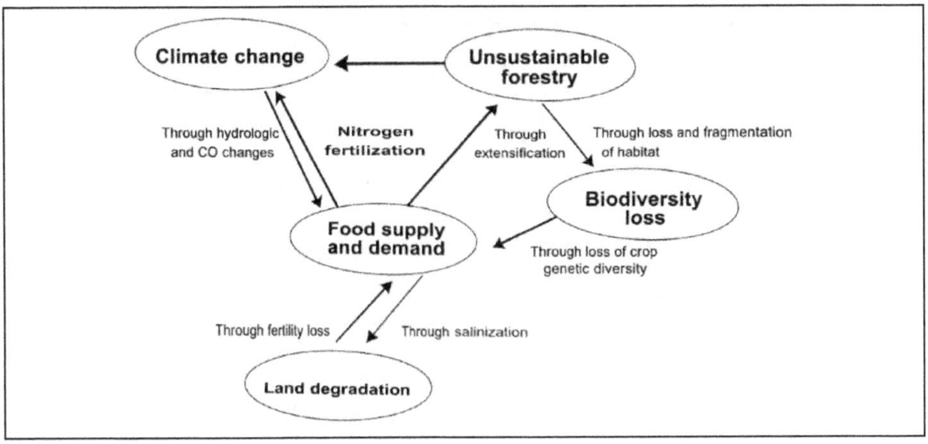

Fig.15.3 Linkages among food production, environmental issues and climate change.
Source: Unknown. Class lecture notes

- "Food supply and demand" is impacted by 'Climate change" through hydrologic and carbon dioxide changes
- "Food supply and demand" is also impacted by "Unsustainable forestry, biodiversity loss, and land degradation"
- All these are interlinked
- Food supply and demand and climate change: Agriculture and associated land use are major sources of greenhouse gases (GHGs), contributing a significant portion of total global emissions
  - o Methane (CH_4): Livestock digestion (enteric fermentation) and manure management are major sources of methane
  - o Nitrous Oxide (N_2O): Fertilizer use and manure management release nitrous oxide
  - o Carbon Dioxide: Deforestation for agriculture and fossil fuel use in farming contribute to carbon dioxide emissions
- Unsustainable forestry and biodiversity loss: Expanding agriculture, particularly into natural landscapes like forests and grasslands contributes to deforestation and fragmentation of habitat, driving biodiversity loss through loss of crop genetic diversity
- Land degradation: Industrial agriculture heavily relies on irrigation and synthetic fertilizers, leading to "Fertility loss and salinization"

15.11 The State of Food Security and Nutrition in the World

- The State of Food Security and Nutrition in the World is an annual report jointly prepared by the Food and Agriculture Organization of the United Nations (FAO), the International Fund for Agricultural Development (IFAD), the United Nations Children's Fund (UNICEF), the UN World Food Program (WFP) and the World Health Organization (WHO)
- Since 1999, the world's progress towards ending hunger, achieving food security and improving nutrition are being monitored and reported regularly. The report also provides an in-depth analysis of key challenges for achieving these goals in the context of the 2030 Agenda for Sustainable Development. The report targets a wide audience, including policymakers, international organizations, academic institutions, and the general public
- The theme of 2025 is "Financing for Development in 2025"

15.11.1 Global Agriculture and Animal Production Statistics

- Overall Agriculture
    - Growing Production: Global agricultural production (including crops, livestock, and aquaculture) has significantly increased over the past six decades, with total output growing nearly fourfold and per capita output increasing by 53 percent
    - Shifting Commodity Mix: There's a notable trend of production shifting from traditional staples like cereal grains and roots/tubers towards a greater emphasis on oil crops, poultry, swine, and aquaculture
    - Major Products: Key agricultural products include foods (cereals, vegetables, fruits, oils, meat, milk, eggs), fiber, fuels, and raw materials
- Animal Production
    - Economic Importance of livestock

  - Livestock contributes significantly to the global economy, representing 40% of the global value of agricultural output and supporting the livelihoods and food security of nearly 1.3 billion people
  - Meat Production: Global meat production exceeded 340 million tons in 2018, representing over 80 billion animals

- Key Livestock Species (2018 estimates)
  - Chickens: 69 billion
  - Pigs: 1.5 billion
  - Cattle: 302 million
  - Sheep: 574 million
  - Goats: 479 million
- Dairy Production
  - In 2021, the world produced 746 million tons of raw cattle milk
- Fish and aquaculture production and aquaculture's ascendancy
  - Aquaculture production of aquatic animals surpassed capture fisheries for the first time in 2022, becoming the main producer
  - Record Production (2022): Global fisheries and aquaculture production: 223.2 million tons,
  - aquatic animals; 185.4 million tons (51% from aquaculture); algae: 37.8 million tons
  - Aquaculture Growth Potential:The growth of aquaculture highlights its potential to meet rising global demand, but sustainable expansion and intensification are needed, particularly in regions like Africa with untapped potential
  - Dominant Producers: A small number of countries dominate aquaculture, with China, Indonesia, India, Vietnam, Bangladesh, the Philippines, Republic of Korea, Norway, Egypt, and Chile accounting for over 89.8 per cent of total production
  - Seafood Consumption: Global consumption of aquatic animal foods reached 162.5 million tons in 2021, driven by factors like rising incomes and urbanization

- Poultry Production (2023)
  - Leading Meat Source: Poultry meat is the most produced meat globally
  - Egg-Laying Hens: China, Indonesia, and India lead global egg production
- Key Trends and Considerations
  - Factory Farming Prevalence: An estimated over 90 per cent of farmed animals globally are raised in factory farms, including 74 per cent of land animals and virtually all farmed fish
  - Biomass Distribution: Livestock now makes up 62 per cent of the world's mammal biomass, highlighting the significant human impact on the mammal kingdom
  - Sustainable Practices: The increasing demand for food necessitates sustainable practices in agriculture, fishing, and aquaculture to address challenges like climate change, resource depletion, and biodiversity loss

15.11.2 Climate Change, Food Supply, and Demand

- Historically, farmers and livestock herders recognized that climate variability has a major influence on the production of farm and livestock produce and several risk mitigating or avoidance strategies have been developed by them to live with the uncertain weather/climate
- However, at present human driven climate change is challenging food demand, protecting natural resources, and environmental quality
- In the long run climate change results in changes in temperature, changed rainfall patterns, shifts in growing seasons, storms, floods, and droughts which risk the food security and livelihoods of farmers
- Climate change is the principal source of fluctuations in food production and prices of agricultural commodities
- The per caput food supply in the developing countries of 2520 calories per day is 25 per cent below that of the developed countries

- More than a billion people in the world earn less than a dollar a day, most of them in developing countries and face chronic malnutrition
- One third of all pre-school children in developing countries (185 million) are malnourished and are underweight of their age
- Experience from 1973, when grain stocks reached a low of 55 days of consumption, showed that when grain carryover stocks fall below 60 days of consumption, markets can become very volatile, easily influenced by weather
- Accelerated industrialization and urbanization consume large amounts of crop land
- Developing countries are projected to increase their cereal demand by about 55 per cent
- A specific climatic event can have a considerable effect on commodity prices. Example: A very hot and dry 1980 season in the United States caused a dramatic rise in prices of soybean. Such weather-induced prices, often coupled with similar price increases in other key commodities, play a major role in the overall price structures
- Around 733 million people faced hunger in 2023, equivalent to one in eleven people globally and one in five in Africa
- The world may be falling significantly short of achieving Sustainable Development Goal (SDG) 2, goal of zero hunger by 2030. The world has been set back 15 years, with levels of undernourishment comparable to those in 2008-2009
- Despite some progress in specific areas such as stunting, an alarming number of people continue to face food insecurity and malnutrition as global hunger levels have plateaued for three consecutive years, with between 713 and 757 million people undernourished in 2023, approximately 152 million more than in 2019 when considering the mid-range (733 million)
- Regional trends vary significantly. The percentage of the population facing hunger continues to rise in Africa (20.4 percent), remains stable in Asia (8.1 percent) though still represents a significant challenge as the region is home to more

than half of those facing hunger worldwide and shows progress in Latin America (6.2 percent). From 2022 to 2023, hunger increased in Western Asia, the Caribbean, and most African subregions

- If current trends continue, about 582 million people will be chronically undernourished in 2030, half of them in Africa, warning the Food and Agriculture Organization of the United Nations (FAO), the International Fund for Agricultural Development (IFAD), the United Nations Children's Fund (UNICEF), the UN World Food Program (WFP), and the World Health Organization (WHO). This projection closely resembles the levels seen in 2015 when the Sustainable Development Goals were adopted, marking a concerning stagnation in progress

15.11.3 Key Findings beyond Hunger

- The access to adequate food remains elusive for billions. In 2023, around 2.33 billion people globally faced moderate or severe food insecurity, a number that has not changed significantly since the sharp upturn in 2020, amid the COVID-19 pandemic. Among those, over 864 million people experienced severe food insecurity, going without food for an entire day or more at times. This number has remained stubbornly high since 2020 and while Latin America shows improvement, broader challenges persist, especially in Africa where 58 percent of the population is moderately or severely insecure
- The lack of economic access to healthy diet also remains a critical issue, affecting over one-third of the global population. Over 2.8 billion people were unable to afford a healthy diet in 2022. This disparity is most pronounced in low-income countries, where 71.5 percent of the population cannot afford a healthy diet, compared to 6.3 percent in high-income countries. Notably, the number dropped below pre-pandemic levels in Asia and in Northern America and Europe, while it increased substantially in Africa

- While progress has been made in increasing exclusive breastfeeding rates among infants to 48%, achieving global nutrition targets will be a challenge. Low birthweight prevalence has stagnated around 15%, and stunting among children under five, while declining to 22.3%, still falls short of achieving targets. Additionally, the prevalence of wasting among children has not seen significant improvement while anaemia in women aged 15 to 49 years has increased
- Similarly, new estimates of adult obesity show a steady increase over the last decade, from 12.1 percent (2012) to 15.8 percent (2022). Projections indicate that by 2030, the world will have more than 1.2 billion obese adults. The double burden of malnutrition the co-existence of undernutrition together with overweight and obesity, has also surged globally across all age groups. Thinness and underweight have declined in the last two decades, while obesity has risen sharply
- These trends underscore the complex challenges of malnutrition in all its forms and the urgent need for targeted interventions as the world is not on track to reach any of the seven global nutrition targets by 2030, the global five agencics indicate
- Food insecurity and malnutrition are worsening due to a combination of factors, including persisting food price inflation that continues to erode economic gains for many people in many countries. Major drivers like conflict, climate change, and economic downturns are becoming more frequent and severe. These issues, along with underlying factors such as unaffordable healthy diet, unhealthy food environments, and persistent inequality are now coinciding, simultaneously, amplifying their individual effects

15.11.4 Financing to End Hunger

- It is emphasized that achieving SDG 2 Zero Hunger requires a multi-faceted approach, including transforming and strengthening agrifood systems, addressing inequalities, and ensuring affordable and accessible healthy diet for all. It calls for increased and more cost-effective financing, with a clear and standardized definition of financing for food security and nutrition

15.12 Impacts of Climate Change on Extreme Events and Food Security

- In addition to the points mentioned above, climate change causes serious implications for food security for billions of people in the world
- The climate change driven inter – annual, monthly and daily distribution of climate variables (temperature, radiation, rainfall/precipitation, water vapor pressure in the air and wind speed) affects several physical, chemical, and biological processes that drive the productivity of agricultural systems
- Changes mainly in temperature and rainfall due to climate change influence food production and the consequent food security
- In vulnerable regions, the yields of most of the crops and animals are sensitive to these changes
- Plants and animals are becoming less productive and disappearing in extreme cases of climate change
- Higher temperatures coupled with decreases in water availability due to reduced rainfall, increased evaporation, and evapotranspiration lead to reductions in crop yield
- The consequences of drought are crop failures, dislocation, and famine
- Extremes of heat and cold, droughts and floods, and various forms of extreme weather affect agricultural systems that depend on food
- In conjunction with other physical, social, political, and economic factors, climate change contributes to vulnerability to economic loss, hunger, famine, and dislocation
- Therefore, adaptation to the adverse effects of climate change through sustainable management practices is crucial, which is simultaneously a daunting task

15.13 Critical issues and Strategies to Cope with Climate Change in Food Production

- The world population continues to grow at an unprecedented rate
- When agriculture was first developed around 12,000 years ago, the world population was 10 million
- From the beginning of Christein era, it reached more than first one billion until the industrial revolution in the eighteenth century
- The second billion took another 100 years
- At present the world population is over 9 billion and projections indicate that by the middle of next century, the world population would grow over 11 billion
- The need to improve living standards in the developing countries put enormous pressure on natural resources
- Forests that once covered more than 35 per cent of the global land surface are shrinking
- Massive deforestation interferes with hydrological cycle and effects rainfall of "Convectional origin"
- Increased population occupies marginal lands which are prone to disasters associated with climate change
- Growth rates of production as well as yield of major food crops have reached a plateau
- In addition, since most of the production increases occurred in a non-sustainable manner, even to maintain the same level of production, more external input is required
- Unfavorable weather and climate due to climate change remain the most frequent cause of crop failure leading to widespread distress and even famine

15.14 Options to Deal with Climate Change Impacts on Food Security

- There is an alarming increase in the number of countries affected by extreme events and disasters, which rose from 28 to 46, mostly in the developing world. These have exacerbated the problems of food security, poverty, and land degradation. Climate change driven extremes are on the rise and they continue to target the world's poorest and least developed nations. The options to deal with climate change impacts on food security are,
    - Promoting and continuing the research to develop, disseminate, and apply weather and climate forecast information
    - Maintain, promote and establish the preparedness strategies and mechanisms for droughts, floods, pests, and diseases
    - Improving the weather station networks
    - Create and establish climate change and food security networks and climate outlook forum
    - Multidisciplinary research on improved germplasm for tolerance to biotic and abiotic stresses, and agronomic interventions to mitigate climate change
    - Initiate and strengthen cooperation among academic, research, outreach, and policy institutions working on impacts of climate change on agriculture at national, regional, and international levels
    - Improve agricultural extension services at all levels
    - Develop innovative financial mechanisms for technical and financial support to mitigate and adapt to climate change
    - Improved agrometeorological databases
    - Improving weather and climate prediction methods and dissemination of warnings
    - Agrometeorologists may make efficient use of the improvements in information and communication technologies to provide timely and efficient agrometeorological information and products to cope with impacts of climate change on agricultural productivity

15.15 The Future of Agricultural Meteorology in the Backdrop of Climate Change

- The future of agricultural meteorology is driven by the urgent need to adapt to climate change and ensure assured global food production and food security
- The science of agrometeorology is evolving beyond simply monitoring weather on agriculture to a more complex, systems-based approach that integrates advanced technology, forecasting, and data analysis to help farmers mitigate weather and climate risks and build resilience
- This involves prioritizing both agricultural profitability and focusing on climate-smart and sustainable practices
- The subject of Agrometeorology shall deal in future with teaching, research and extension on issues pertaining to development and dissemination of daily weather, services, and products, early warnings, agromet advisories, and climate resilient agricultural practices in synergy with contemporary developments in computer sciences
- Also, the subject should focus on developing micrometeorological technics to effectively address issues of climate change on agriculture and livestock as the weather and climate can't be modified at meso and macro levels
- The subject shall also emphasize issues viz., use of weather, climate and agrometeorological knowledge as non-monetary inputs in agriculture at a time the cost of cultivation of crops and livestock rearing are becoming beyond the reach of farmers, reduced yields, and quality of produce, more so under climate change scenarios
- New and improved data collection and processing technologies that are transforming agricultural meteorology
    - Remote sensing: Satellite-based sensors and drones provide detailed, real-time data on weather health, soil moisture, crop health, and vegetation cover at a high resolution. This offers critical information for large-scale and precision farming applications

  - Agrometeorological networks: National and international organizations are strengthening data collection networks, including weather stations, to monitor and research on the interactions between weather/climate and agricultural systems
  - Geographic Information Systems (GIS): GIS tools are being integrated with remote sensing to map and analyze agroecological zones, which aids in sustainable management of farming, forestry, and livestock systems
  - Internet of Things (IoT) and Precision Agriculture: On-site monitoring networks are growing in popularity, allowing farmers to use real-time data from their fields to improve efficiency with water and other resources
- Improved forecasting and early warning systems
  - The ability to forecast and provide timely warnings for extreme weather events is a key function of agrometeorology
  - With more erratic weather patterns, the demand for reliable and highly localized agroadvisories will increase
  - The prosperity of agriculture and livestock will depend on improving these services and communicating actionable information to farmers in a timely and accessible manner, often through mobile platforms
  - The accuracy and range of agrometeorological forecasting is expanding, moving from short-term predictions to long-term climate projections
- Seasonal to inter-annual forecasts: These longer-range forecasts help farmers and policymakers plan for potential droughts, floods, and other weather and climate variability
- Extreme weather alerts: Enhanced monitoring systems provide early warnings for extreme events like heatwaves, heavy rainfall, and frost, helping farmers protect crops and livestock
- Climate-smart mitigation and adaptation strategies
  - Agricultural meteorology is now focused on developing strategies to help the agricultural sector mitigate and adapt to climate change

  - Mitigation strategies: Agrometeorologists are researching how agriculture can reduce its own greenhouse gas emissions by adopting practices like conservation village, developing bio-based energy crops, etc.
  - Adaptation strategies: This includes identifying optimal planting dates, using drought-resistant crop varieties, improving irrigation management, shifting to more suitable crops, etc.
- Global collaboration and research
  - International collaboration and knowledge-sharing are growing, linking institutions in developed nations with those in developing climate-vulnerable regions
  - This ensures that research is applicable to local conditions and that solutions are accessible to a wider population, including smallholder farmers
  - Bibliometric studies show an increasing number of publications and partnerships focused on strengthening agrometeorological services globally
- Transition from a reactive discipline to a predictive and strategic science
  - The future of agricultural meteorology is also being reshaped by the rapid development of high-end computer modeling and huge data analysis, transitioning the field from a reactive discipline to a predictive and strategic science. Modern technologies are enabling agricultural meteorologists to provide more precise, localized, and actionable insights that address the immense challenge of global food production and security in an era of climate change
  - The combined effect of quantum computing and Artificial Intelligence (AI) on the future of agrometeorology is expected to be transformational, especially in terms of prediction accuracy, decision making speed, and complexity management. The potential impacts of quantum computing and Artificial Intelligence are
    - Hyper accurate weather and climate forecasting: Quantum computing can handle massive data sets and simulate atmospheric dynamics at granular quantum level, which even the classical models can't manage

effectively. Quantum computing combined with AI could improve (a) local scale weather prediction (down to farm level) (b) long range seasonal forecasting with reduced uncertainty (c) real time extreme weather event alerts (droughts, floods, cyclones, heat and cold waves, etc.)

- Precision agriculture support: Quantum enhanced AI can optimize (a) irrigation scheduling based on forecasts at micro level (b) crop planting windows for maximum yields (c) pest and disease forecasting using weather patterns and quantum simulated behavior of pests and diseases
- Data fusion and complex system modelling: Agromet involves data from soil, water, air, crops, and livestock. The quantum- AI systems can (a) simultaneously model multidimensional interactions (ocean, atmosphere, land biosphere) (b) handle chaotic variables and nonlinear interactions better than traditional models (c) integrate satellite, IoT, drone, and sensor data for a unified agromet model
- Real time decision making for farmers: AI powered agromet advisory systems will help in (a) enhanced quantum optimization for input use (fertilizers, water for irrigation, etc.) (b) provide tailored weather-based advisories with real time risk assessments (c) enable multi language region specific alerts with voice or video AI interfaces
- Resilience against climate change: Quantum- AI models will enable (a) scenario simulations under climate stress (combined effect of temperature rise in rainfall) (b) early warning systems for food security, (c) crop- climate modelling for new varieties
- Challenges ahead: (a) high cost and limited availability of quantum computing resources (still in early development) (b) need for large and labeled datasets to train AI effectively (c) interdisciplinary collaboration is required for agrometeorologists with all agricultural, physics, computer sciences, social sciences, etc.(d) accuracy under extreme or rare weather conditions

- o The next generation agricultural Meteorology will be more predictive, personalized, and climate resilient

15.16 Key Trends in Agricultural Meteorology

- High-resolution, data-driven forecasting is a new trend in agrometeorology
- While traditional numerical weather prediction (NWP) models require massive computational power, new data-driven models are dramatically changing the field
    - o AI and Machine Learning: AI and deep learning models can produce localized forecasts in seconds for specific geographical areas for agriculture
    - o Prediction of hyper-local climate at micro level: The combination of IoT sensors, satellite imagery, and on-farm weather stations allows AI to generate predictions not just for a region, but for specific fields. This allows farmers to get granular data on expected temperature, soil moisture, and rainfall
- Enhanced predictive and prescriptive analytics
    - o Modern meteorological analysis moves beyond just predicting weather to prescribing specific actions for farmers
        - ▪ Optimized resource management: By analyzing data from sensors and weather models, AI algorithms can provide farmers with precise recommendations for irrigation, fertilization, and pesticide application. This reduces waste and minimizes environmental impact
        - ▪ Early warning systems: AI-powered systems can detect early signs of pest infestations, crop diseases, and drought stress by cross-analyzing sensor data, satellite imagery, and weather forecasts
        - ▪ Risk mitigation: Predictive models help farmers assess the risk of crop damage from adverse weather events like heatwaves, heavy rainfall, and frost informing decisions on crop selection, planting schedules, and field operations

- Crop resilience and genomics
  - Agricultural meteorology is now influencing the development of hardier and climate-resilient crop varieties
    - Climate-resilient crops: AI models analyze vast genetic and environmental datasets to help breeders identify genetic traits linked to drought tolerance, heat resistance, and other climate-resilient characteristics
    - Precision breeding: By integrating high-throughput phenotyping with genomic data, AI can expedite the development of new plant types optimized for changing climatic conditions
- Supply chain optimization and policy
  - The insights from advanced agrometeorological models are moving beyond the farm to affect the broader agricultural ecosystem
    - Market trend analysis: Predictive analytics inform agribusinesses and policymakers about potential crop yields, supply chain disruptions, and market fluctuations
    - Data-driven policy: Governments are increasingly using AI insights to formulate agricultural policies, design insurance plans, and manage disaster response programs
    - Sustainable development: By optimizing resource use and mitigating climate risks, agricultural meteorology supports green development and eco-friendly farming practices

15.16.1 Challenges and Opportunities

- Challenges
  - The "last mile" barrier
    - A significant challenge is ensuring that localized, actionable information reaches all end-users, especially marginalized farming communities. This requires overcoming issues like inadequate infrastructure, limited financial resources, and poor internet connectivity, particularly in developing countries

- Integrating Indigenous knowledge
  - Successful adaptation will require integrating indigenous and traditional agricultural knowledge with modern scientific models. This process is complex and demands developing frameworks that build trust and effectively combine traditional insights with scientific methodology
- Addressing farmer needs: A "bottom-up" approach that involves farmers in developing services is crucial to ensure information is relevant, accessible, and applicable to their challenges
- Data integration and quality: Large-scale implementation is hindered by fragmented systems, inconsistent data quality, and a lack of data interoperability
- High costs and accessibility: The costs associated with acquiring, processing, and interpreting large datasets can be prohibitive for small and medium-sized farms
- Model limitations: AI models may lack interpretability and face challenges in evaluating uncertainty and predicting rare and extreme weather events
- Digital divide: In developing nations, the lack of computational power, meteorological expertise, and rural bandwidth can limit the adoption of advanced agrometeorological tools

- Opportunities
  - Collaboration and data sharing: International and public-private partnerships can foster data sharing and innovation to address global agricultural challenges
  - Democratization of technology: The development of cloud-based, mobile-friendly platforms, and open-source models is making advanced agrometeorology accessible to a broader user base
  - Explainable AI: Advancements in explainable AI will increase trust and understanding in complex predictive models, facilitating their adoption by farmers and stakeholders

- o Future workforce: Cultivating a new generation of agricultural meteorologists proficient in both agricultural meteorology and data analytics is crucial for advancing the field

15.17 Role of Agrometeorologists for Profitable Agriculture, Food Production, and Food Security under Climate Change

- Teaching, research, and extension
  - o Agricultural meteorologists play a vital role in developing sustainable and climate-resilient agricultural systems. Under climate change, their duties extend beyond traditional forecasting for building resilience
  - o The Agrometeorologists shall deal with teaching, research, and extension on issues pertaining to value added agro advisories, development and dissemination of daily weather, services, and products, early warnings, and climate resilient agricultural practices
  - o Also, shall develop micrometeorological technics to solve the micro level negative impacts of climate change on agriculture and livestock as weather/climate can't be modified at meso and macro levels
  - o They shall also focus on issues such as use of weather, climate and agrometeorological knowledge as non-monetary inputs under climate change scenarios for prosperity in the agriculture production systems. The reason is that the cost of cultivation of crops and animal rearing is alarmingly increasing and it is beyond the affordability of farmers. Yields and quality of produce are reducing. Climate change is adding fire to fuel
- Providing actionable climate information
  - o Climate and weather forecasting: Agricultural meteorologists provide essential information on weather and climate from seasonal predictions to real-time to short-term

forecasts. This allows farmers to make informed decisions on planting/sowing, harvesting, and other daily operations

- o Monitoring and analysis: They constantly monitor and analyze climate and agronomic variables, using modern tools like Remote Sensing and GIS to provide spatial data. This is essential for assessing climate impacts and tracking greenhouse gas (GHG) emissions from the agricultural sector

- Building resilience and adaptation
  - o Supporting climate-smart agriculture (CSA): Agrometeorologists are central to implementation of CSA by providing the data needed to enhance productivity and increase resilience
  - o Advising on farming practices: They offer advisory services that enable farmers to adapt to changing climatic conditions. This includes advice on managing water and fertilizer use, controlling pests and diseases, and science-based sowing and harvesting schedules
  - o Promoting resilient practices: They promote adaptation practices, including using resilient crop varieties and locally adapted species to build the resilience of production systems and livelihoods
- Enhancing early warning systems
  - o Contributing to multi-hazard early warnings: The WMO leads the forecasting component of the global "Early Warnings for All" (EW4All) initiative, which aims to protect everyone from hazardous weather. FAO contributes by tailoring these systems to agricultural communities and implementing anticipatory actions. Agrometeorologists play significant role in implementing this initiative and take it to farmers' field level
- Capacity building: A key duty is strengthening the ability of all stakeholders to provide meteorological services to their agricultural communities

- Informing adaptation plans: Agrometeorologists ensure that national adaptation plans (NAPs) incorporate relevant, actionable, and accessible climate information for agricultural communities. This links climate data to policy development and implementation
- Knowledge sharing: Agrometeorologists help translate technical meteorological information into practical, usable knowledge for farmers, agricultural extension workers, and policymakers. They also bridge the gap between global forecasts and the specific needs of local communities
- In future, agrometeorologists will evolve from data providers to interdisciplinary problem-solvers. They will work with a range of experts, from weather and climate modelers to sociologists, to develop sustainable and resilient agricultural systems. Their role will be to
    - Integrate diverse data sources and improve forecasting models
    - Translate complex scientific data into practical, actionable advice for farmers
    - Collaborate with policymakers to develop climate-resilient agricultural policies
    - Conduct agrometeorological, micrometeorological, and multidisciplinary research to understand the complex interactions between climate, biodiversity, and agricultural production
    - Strengthen educational and outreach programs to help farmers adapt to changing climate conditions
    - They play an increasingly vital role in securing global food supplies and helping the agricultural sector adapt to the changing climate

15.18 Addressing the Climate Change Challenges on Food Production and Food Security

- To address climate change challenges, individuals, families, communities, governments, organizations etc., have the best sciences available to plan for and implement climate-smart and resilient practices. Two critically important approaches which are available to managing climate resilience are mitigation and adaptation. These strategies are interrelated and must be pursued simultaneously as both can potentially improve resilience to the changing climate.

Chapter 16

Global Conventions, Treaties, Summits, and Schools of Thought on Climate Change

"Global conventions and summits, guided by Vedic dharma, should strive to foster the principle of "Vasudhaiva Kutumbakam" – the world as one family. They aim to promote harmony, ethical conduct (including truth and non-violence) and the collective well-being of all beings. These gatherings should act for the welfare of the world (loka Sangraha) working towards sustainable and ethical solutions for global challenges"

16.1 Introduction to Global Conventions, Treaties, and Summits on Climate Change and Agriculture

- Greenhouse gas emissions from one nation contribute to global warming that impacts the entire world. The Earth's atmosphere, oceans, and ecosystems are interconnected, and the effects of climate change are felt globally and require a unified and international response, not a collection of isolated national actions

- Addressing climate change requires an unprecedented level of international cooperation to foster emission reductions and technological innovation. No single nation, even a large emitter, can solve the crisis independently. Global agreements create a framework for countries to work together, set goals, share resources, and monitor progress
- Effective climate action demands significant financial investment, technological innovation, and scientific expertise. No single entity possesses all these resources. Global collaborations facilitate the sharing of green technology, research, best practices, and financial assistance particularly between developed and developing nations. Example: The Montreal Protocol successfully led to a significant reduction in ozone-depleting substances, showcasing the power of collective commitments to environmental health
- Climate change disproportionately affects vulnerable populations and developing countries, even though they have historically contributed the least to the problem. International cooperation, through mechanisms like the Loss and Damage Fund and climate finance, aims to provide financial support and expertise to help these nations adapt and build resilience
- International climate agreements often establish frameworks for countries to submit and regularly update Nationally Determined Contributions (NDCs), detailing their plans to reduce greenhouse gas emissions and enhance climate resilience. This process allows for monitoring and reporting on progress, fostering transparency, and accountability in reaching shared climate goals. For instance, the Paris Agreement encourages nations to increase their climate ambitions every five years
- Environmental degradation caused by climate change can lead to conflicts over resources, displacement, and economic instability. International cooperation promotes sustainable resource management, reducing the risk of environmental-driven conflicts and strengthening diplomatic ties. It also presents economic opportunities, including fostering investment in clean energy technologies and creating jobs in green sectors

- In essence, global conventions and summits provide a platform for nations to come together, acknowledge the shared threat of climate change, and work collaboratively towards a more sustainable and resilient future for all

16.2 Important Global Conventions, Treaties, and Summits on Climate Change and Agriculture

- Since 1987, six (6) global environmental treaties have been signed which signifies the importance accorded to environment and climate Change in international relations. Among these, three (3) Conventions are significant to the subject of climate change, food production, and food security. Since 1992, the world community has negotiated and ratified these three important International Conventions, all of which have a significant bearing on climate change and sustainable agriculture as well. These are
    - The United Nations Framework Convention on Climate Change (UNFCCC)
    - The Convention on Biological diversity (CBD)
    - United Nations Convention to Combat Desertification (UNCCD)
- In addition to the above, the "World Food Summit Plan of Action (WFSPA)" was developed in 1996, which included several commitments by international community to make agricultural production sustainable, even under climate change

16.3 The United Nations Framework Convention on Climate Change (UNFCCC)

- The UNFCCC was adopted in May 1992 and opened for signature at the 1992 Earth Summit in Rio de Janeiro. It entered into force in March 1994 and as of May 2018 had 197 Parties (196 States and the European Union). The Convention's ultimate objective is the "Stabilization of greenhouse gas concentrations in the atmosphere at a level that would prevent dangerous anthropogenic interference with the climate system". The provisions of the Convention are pursued and implemented by two treaties: the Kyoto Protocol and the Paris Agreement

- This Convention highlights several points dealt with by international organizations and the need for international response. This applies to climate where due to the very nature of the atmosphere, all physical causes and effects are global in nature. The Convention also stressed on three issues, viz.,
    - The access to resources to achieve sustainable development
        - The ultimate objective of the Convention is to achieve stabilization of greenhouse gas concentrations at a level that would prevent dangerous anthropogenic interference with the climate system
        - Such a level should be achieved within a time frame sufficient to allow ecosystems to adapt to climate change, to ensure that food production is not threatened, and to enable economic development to proceed in a sustainable manner
        - Agriculture has the potential to adapt much faster than other sectors essentially in field crops, livestock rearing, inland fisheries, and animal production
        - Adaptation may be more difficult in sectors such as grazing in natural pastures, ocean fisheries, etc.
        - It was also mentioned to focus on policies to protect climate systems integrated with national development programs, by taking all sources and sinks into consideration
    - The recognition of basic differences between developed and developing countries
        - Agriculture is mentioned as both sources and sinks of greenhouse gases. The database available through research and systematic observation in agriculture is significantly poorer in developing countries than in developed countries

        - The standards of methodological observations in agriculture are less in developing than in developed countries
        - Agriculture is at subsistence level and Governments pay little attention to agriculture as compared to other sectors like energy in developing countries than the developed countries
    - At the same time due to population pressure and urbanization, many farmers are currently forced out of their traditional subsistence systems into more productive and market-oriented types of farming
        - This results in changing local and global environments because of new crop varieties and technics that use more synthetic fertilizers
        - In developed countries the farm holdings are huge in size. On the other hand, from farm size wise the farmers of developing countries are small and marginal
    - The major commitment
        - In this Convention the major commitment was to promote and cooperate in research, systematic observation, and development of data archives related to climate systems

16.4 Green Climate Fund

- The Green Climate Fund (GCF) is the world's largest dedicated climate fund, created under the United Nations Framework Convention on Climate Change (UNFCCC). Its mandate is to help developing countries finance climate action
- Mission
    - The GCF aims to achieve a "Paradigm shift" toward low-emission and climate-resilient development by investing equally in mitigation and adaptation projects. It primarily serves nations most vulnerable to climate impacts, including Least Developed Countries (LDCs) and Small Island Developing States (SIDS)

- Operations
    - Funding: The fund receives pledges from developed countries and mobilizes finance from private sources to fund projects
    - Approval Process: Developing countries propose projects and access financing through accredited national, regional, and international partner organizations
    - Financing tools: Funding is provided through a mix of grants, loans, and equity investments
- Noteworthy achievements
    - Since approving its first projects in 2015, GCF has grown to become the largest multilateral climate fund, with a portfolio of billions of dollars
    - It has funded hundreds of mitigation and adaptation projects in over 100 countries
    - In 2025, the GCF announced its expansion with the launch of regional offices

16.5 The KYOTO Protocol

- The third Conference of Parties (COPs) of the UNFCCC which took place in Kyoto, Japan, during the late 1997 led to the "Kyoto Protocol"
- The Kyoto protocol, linked to UNFCCC, became effective on February 6th, 2005
- This protocol focused explicitly on sustainable management of agriculture and forestry in the light of climate change
- The Kyoto Protocol commits the signatories to reduce overall emissions of greenhouse gases by 5 per cent below the 1990 levels in the commitment period from 2008 to 2012
- More specifically enteric fermentation, manure management, rice cultivation, agricultural soils, and burning of savannahs and field burning of agricultural residues are all listed as sources which will have to be taken into consideration by countries in their reports to the Conference of Parties (COPs) of anthropogenic sources and sinks

- The Kyoto Protocol also makes provision for Clean Development Programs (CDP) to assist the developing Parties with dual purposes (a) to achieve sustainable development (b) compliance with emission limitations and reduction commitments
- The Clean Development Mechanism will allow Parties to jointly implement projects, which will result in certified emission limitations. This mechanism complements the tradable permit approach
- The implementation of both Tradable Permits and the Clean Development Mechanism will require a detailed inventory of sources and sinks, regular reporting as well as review and control mechanisms, which certainly require improved methodologies to assess emissions and sinks
- The Protocol emphasizes on "Environmentally sound technologies" in the concept of "Sustainability"
- There are also commitments in the Protocol for countries to formulate national and regional programs to improve scientific cooperation, education and training, inventories of emissions and sinks, and adaptation potential
- It was also insisted that the developed Parties shall assist developing Parties
- India ratified the Protocol
- Compared with the UNFCCC, the Kyoto Protocol focuses much more explicitly on agriculture and forestry on the "Promotion of sustainable forms of agriculture in the light of climate change considerations and on sustainable forest management practices"

16.6 The Paris Agreement

- The Paris Agreement is a "Legally binding international treaty on climate change"
- To tackle climate change and its negative impacts, world leaders at COP 21 in Paris organized by the UNFCCC reached a breakthrough on 12 December 2015 in which 192 parties have joined

- It entered in to force on 4 th November 2016. The points stressed in the historic Paris Agreement are
- To strengthen the global climate change response by increasing the ability of all participating nations to adapt to adverse impacts of climate change and foster climate resilience
- The Agreement sets long term goals to guide all nations to
  - Substantially reduce global greenhouse gas emissions to limit the global temperature increase in 2000-2099 to 2 degrees centigrade while pursuing the efforts to limit the increase even further to 1.5 degrees centigrade
  - Review the commitments of all the countries every five years
  - Provide financing to developing countries to mitigate climate change, strengthen resilience, and enhance abilities to adapt to climate change impacts
  - Each Nation shall design its own emissions reduction plan, known as NDC
  - These plans are revisited and strengthened periodically, fostering a system of continuous improvement over time
  - The Agreement requires all nations to engage in adaptation planning and implementation through
    - National adaptation plans
    - Vulnerability assessments
    - Monitoring and evaluation
    - Economic diversification
    - Communicate priorities, plans, actions, and support needs through adaptation mechanisms
- These shall be recorded in a public registry

16.7 The Conference of Parties 26 (COP 26)

- The Twenty-sixth session of the Conference of the Parties (COP 26) of the United Nations Framework Convention on Climate Change (UNFCCC) was held in Glasgow, UK from 31 October to 12 November 2021. Projections from the Twenty-sixth session of the Conference of the Parties (COP 26) of UNFCCC are

- The COP26 summit brought the Parties together to accelerate action towards the goals of the Paris Agreement and the UNFCCC
- For starters, it recognizes the global climate emergency, citing recent IPCC findings
- It expresses “Alarm and utmost concern that human activities have caused around 1.1°C of global warming by 2021 and that the impacts are already being felt in every region”
- Parties also recognize that the impacts of climate change will be much lower at a 1.5°C rise compared with 2°C, and resolved to pursue efforts to keep to 1.5°C
- The Conference significantly ramps up the call for greater action and financing for adaptation
- It urges developed countries to at least double their collective climate finance for adaptation in developing countries from 2019 levels by 2025, to ensure a balance between mitigation and adaptation
- It calls on multilateral development banks, other financial institutions and the private sector to enhance finance mobilization to deliver the scale of resources needed to achieve climate plans
- Another focus in COP 26 was the need for more lending to help poor countries shift out of fossil energy, to build societies that are better adapted to the impacts of climate change and to recover from the damage caused by extreme weather events
- In a statement on the conclusion of COP26, the UN Secretary-General António Guterres said the conference outcome reflected the state of political will in the world and provided building blocks for progress. “We are in the fight of our lives, and this fight must be won,” he said. “Never give up. Never retreat. Keep pushing forward”

16.8 The Conference of Parties 29 (COP29)

- The UN Climate Change Conference (COP29) organized in November, 2024, at Baku, Azerbaijan with a new finance goal to help countries to protect their people and economies against climate disasters, and share in the vast benefits of the clean energy boom. With a central focus on climate finance, COP29 brought together nearly 200 countries, and reached a breakthrough agreement that will,
    - Triple finance to developing countries, from the previous goal of USD 100 billion annually, to USD 300 billion annually by 2035
    - Secure efforts of all actors to work together to scale up finance to developing countries, from public and private sources, to the amount of USD 1.3 trillion per year by 2035
- Known formally as the New Collective Quantified Goal on Climate Finance (NCQG), it was agreed after two weeks of intensive negotiations and several years of preparatory work, in a process that requires all nations to unanimously agree on every word of the agreement
- This new finance goal is an insurance policy for humanity, amid worsening climate impacts hitting every country,". "But like any insurance policy – it only works – if premiums are paid in full, and on time. Promises must be kept, to protect billions of lives"
- It will keep the clean energy boom growing, helping all countries to share in its huge benefits: more jobs, stronger growth, cheaper and cleaner energy for all
- The International Energy Agency expects global clean energy investment to exceed USD 2 trillion for the first time in 2024
- The new finance goal at COP29 builds on significant strides forward on global climate action at COP27, which agreed an historic Loss and Damage Fund, and COP28, which delivered a global agreement to transition away from all fossil fuels in energy systems swiftly and fairly, triple renewable energy and boost climate resilience

- COP29 also reached agreement on carbon markets – which several previous COPs had not been able to achieve. These agreements will help countries deliver their climate plans more quickly and cheaply, and make faster progress in halving global emissions this decade, as required by science

16.8.1 Adaptation

- COP29 was an important moment for adaptation, with the delivery of several key outcomes. The COP decision on matters relating to the least developed countries (LDCs) contains a provision for the establishment of a support programme for the implementation of National Adaptation Plans (NAPs) for the LDCs. Parties extensively discussed the second five-year assessment of progress to formulate and implement NAPs
- A High-Level Dialogue on National Adaptation Plans convened ministers from least developed countries and small island developing States, financial experts and international donors to address the growing urgency of climate adaptation. Their discussions focused on innovative financing, technical support, and accelerated action to meet the 2025 submission deadline for NAPs. The event concluded with a strong call to action to expedite NAPs and translate plans into tangible outcomes
- The outcome on the global goal on adaptation sets a clear path forward on the road to COP30 for the indicators work programme, providing a process for experts to continue their technical work before passing the baton to Parties. COP29 also launched the Baku Adaptation Road Map and Baku high-level dialogue on adaptation to enhance the implementation of the UAE Framework. Finally, the outcome raises ambition by agreeing to continue unpacking transformational adaptation moving forward
- COP29 took a decisive step forward to elevate the voices of Indigenous People and local communities in climate action, adopting the Baku Workplan and renewing the mandate of the Facilitative Working Group (FWG) of the Local Communities

and Indigenous Peoples Platform (LCIPP). The adopted decision acknowledges the progress made by the FWG in fostering collaboration among Parties, Indigenous People and local communities, and underscores the leadership of Indigenous People and local communities in addressing the climate crisis

16.8.2 Gender and Climate Change

- Countries agreed a decision on gender and climate change, extending the enhanced Lima Work Program on Gender and Climate Change for another 10 years, reaffirming the importance of gender equality and advancing gender mainstreaming throughout the convention

16.9 The Convention on Biological Diversity (CBD)

- The Convention on Biological Diversity (CBD) which is based on a broad ecosystem approach was opened for signatures at the Earth Summit in Rio de Janerio, Brazil, on 5th June 1992. It entered into force on 29th December 1993. The Convention contains three national level obligations
    - Conserve biological diversity
    - Sustainably use biological diversity
    - Share its benefits
- Every cubic centimeter of the biosphere has been altered by human-induced changes in the climate and the chemical composition of the atmosphere, and this carries major implications for biological diversity
- Together plants, animals, and microbes absorb and break down pollutants, help maintain a mix of gases in the atmosphere, regulate the solar energy the earth absorbs, moderate regional weather, modulate the water cycle, minimize floods and drought, and purify waters
- Hence, there are major interactions between climate and biological diversity
- The current decline in biological diversity is largely the result of human activity. Despite mounting efforts over the past two

decades, the loss of world biological diversity is mainly due to habitat destruction, overharvesting, pollution, and the inappropriate introduction of foreign plants and animals

- Forest ecosystems are vital for biological diversity. They provide the most diverse sets of habitats for plants, animals, and microorganisms
- Forests and forest biological diversity play important role in the lives of many indigenous and local communities
- Deforestation disturbs the dynamic coupling between forest and atmosphere that sustains unique regional climate and biological diversity
- Forests hold most of the land's carbon stores
- A better understanding of the climate of the major ecosystems of the world where biological diversity is at risk could help develop effective *in situ* conservation strategies
- Any natural resource conservation measures to maintain land productivity through sustainable agricultural practices could effectively contribute to the maintenance of biological diversity

16.10 United Nations Convention to Combat Desertification (UNCCD)

- The "Desertification" and "Dryland degradation" are synonymous
- The total area of arid, semi- arid, and dry sub- humid drylands cover 40 per cent of earth's land surface
- Desertification was recognized as a major worldwide issue following the disastrous droughts in the Sahel during the early 1970s
- This led to the United Nations Conference on Desertification (UNCOD) in Nairobi in 1977, which formulated the plan of action to combat desertification
- An important outcome of the Earth Summit in 1992 was an agreement to draw up an international Convention to combat desertification, which was elaborated by the "Inter-governmental Negotiating Committee on Desertification (INCD)" and finally agreed in June, 1994

- The United Nations Convention to Combat Desertification (UNCCD) entered into force on 26th December 1996 and several countries ratified. It recognizes that the desertification issue is on a par with other global issues such as climate change and loss of biodiversity
- The following areas identified in the Convention are of specific interest to agrometeorology and climate change
    - Incorporation of long-term strategies to combat desertification and mitigate the effects of drought in the national action programs
    - Enhancing national climatological, meteorological, and hydrological capabilities and the means to provide for drought early warning
    - Establishment and or strengthening early warning systems
    - Strengthening drought preparedness and management strategies
    - Development of sustainable irrigation programs not only for crops but also to make availability of water to animals
    - Strengthening capacity building, education, and public awareness on meteorological and hydrological services
    - Early warning and advance planning for climate change
    - Ensuring that the collection, analysis, and exchange of information addresses the needs of local communities and decision makers
    - Contribute to knowledge of processes leading to desertification and drought
    - Making inventories of relevant traditional and local technology
    - Encouraging and supporting the development of new technologies
    - Training of decision makers, managers, and personnel working on early warning systems on drought for food production

- This Convention calls for enhancing national climatological, meteorological, and hydrological capabilities and the means to provide for drought early warning
- The national action programs under the UNCCD are required to strengthen drought preparedness and management, including drought contingency plans which take into consideration seasonal and inter annual climate predictions
- Much of the responsibility to help the local populations in coping with the consequences of droughts, floods etc., lies with the National Governments. Pro- active action, rather than "Wait and see" attitude on the part of the Governments can often help avoid a serious drought turning into a famine
- Although there were major droughts and climate induced food shortages around the globe in the recent years in various parts of Africa, India, China, Brazil, and Indonesia, famines occurred only in Africa
- What it implies is that *ad hoc measures* of drought relief should be replaced by more concerted, systemic long term action programs that provide sustained support and advise to the local populations to develop their ability to cope with the climate change

16.11 The World Food Summit (WFS)

- The World Food Summit (WFS) was hosted by FAO, Rome in November 1996
- The main objective was to improve food security at all levels and to significantly reduce the number of undernourished people
- The Summit recognized "Poverty" as the main factor for food insecurity. The Summit also recognized that desertification and loss of biodiversity are due to climate change and human interference
- Therefore, the Summit emphasized and suggested to maintain, promote, and establish preparedness strategies and mechanisms like development and application of climate forecast information for surveillance and early warning of droughts, floods, and pest and disease alertness

- The food needs of future populations have to be met from increasing the productivity per hectare
- At the Summit several commitments were negotiated to ensure that "All people, at all times, have physical and economic access to sufficient, safe, and nutritious food"
- In this Summit and the "World Food Summit Plan of Action" it was suggested to incorporate objectives that make explicit reference to the impact of climate variability/ change on food security
- The suggestions of the "WFS Plan of Action" for food security due to climate change are
    - Identify the potential and improve the productive use of national land and water resources for sustainable increases in food production
    - Better understanding of impacts of global environmental threats, in particular climate change/ climate variability
    - Promote and continue international research on climate forecasts and their dissemination
    - Maintain, promote, and establish collaborations with NGOs for development and application of climate forecast information for surveillance and early warning, drought, flood, other natural disasters, and pest and disease alertness
    - Support international efforts to develop and apply climate forecast information to improve the effectiveness and efficiency of emergency preparedness
- As stated above in the World Food Summit Plan of Action, the Science of Meteorology and Agricultural Meteorology have very important roles to play in developing effective strategies for coping with climate change in general and agriculture in particular

16.12 Introduction to Schools of Thought on Climate Change

- The phenomenon of climate change is present in the very rotation of the earth around itself with an inclined axis

- With each rotation the position of the axis of the earth shifts infinitesimally and imperceptibly
- Accumulated over many centuries, such a shift can lead to reversal of agricultural seasons locally
- The shift of local seasonal features with time has been noticed in ancient times
- The ancients believed that the seasonal features would revert and termed the phenomenon as "The Oscillation of Equinoxes"
- Newton opined that the observed changes in seasons were unidirectional and termed it "The Procession of Equinoxes"
- It is estimated that a full reversal of season from Winter to Summer and vice versa will occur locally once in 45000 years plus or minus 5000 years
- There are reports of the Tropics of Cancer shifting southwards at about 1.27 km per 100 years
- Due to normal shifts in Earth's obliquity and external gravitational influences, this will lead to shrinking of the tropics and setting in of harsher climates in the current tropical areas
- Again, for some unknown reason the earth gets covered once every one lakh years by ice and this gives rise to inter-glacial periods of warm and ice ages
- Highly perceptible changes in climate can occur locally even over a few centuries
- Anthropogenic activities have brought, within the realm of reality, the possibility of perceptible climate changes even within decades

16.12.1 Schools of Thought on Agriculture

- During late 1990s and early 2000s there were two opposing schools of view in the field of climate change in general and in the field of agriculture in particular advocating cognizance and ignoring climate change. Some of the arguments advanced by both schools of view deserved merit consideration at that time. This dichotomy arises from the fact that

- A change in each weather element does not occur in isolation but is reflected in changes of other weather elements as well
- The effect of combination of changes on any physical or biotic entity/process is quite different from what would result for a given change in a weather element
- The spatial variations of climate on earth are due to the coverage of the planet with different types of surfaces, each with its own radiative temperature even under the same input of energy
- Thus, any large-scale change in the type of surface, like deforestation, will bring about local climate change
- Again, pollution of soil and water resources brings about a dwindling down of aquatic fauna and terrestrial flora
- For any weather parameter, there are decades of epochs of increase and decrease, which will affect biotic processes in a manner like that of climate change
- However, notwithstanding the above there is enough evidence by way of increased surface temperatures over base period, melting polar ice caps, retreat of glaciers etc., to confirm that global warming and climate change are taking place. It is due to injection of large volumes of GHGs and their interference in the energy balance of the earth-atmosphere system and warming of the earth's surface
- Climate change is further exacerbating the natural weather vagaries and effects of anthropogenic follies and resulting in a significant and visible impact on crops and livestock and affecting the whole agriculture to a large extent
- It is suggested that for assessing impact of climate change on crops and livestock it is better to take, on a region-wise and season-wise basis, recently observed rate of change of parameters for the business-as-usual climatic scenario, twice the current rate of change for the pessimistic climatic scenario and 50 per cent of the current rate for the most optimistic scenario

16.12.2 Schools of Thought on Afforestation and Reforestation

- Forests provide immense benefits by removing carbon dioxide and pollutants from the atmosphere, preventing soil erosion, filtering water, and housing half of the world's land species of animals, plants and insects. Reforestation and afforestation are two of the most effective nature-based solutions in fighting climate change and limiting its impacts

- Reforestation is the process of replanting trees in areas that had recent tree cover but where forests were lost, due to wildfires, drought, disease, or human activity such as agricultural clearing
- Afforestation is the process of planting trees in areas that have not been forested in recent history. Afforestation helps restore abandoned and degraded agricultural lands, prevent desertification, create carbon sinks, and generate new economic opportunities for local communities
- Afforestation and reforestation help mitigate climate change by sequestering carbon dioxide through photosynthesis, while also improving adaptation by enhancing biodiversity, water quality, and providing local safety nets. This is most essential under global warming
- Role in Climate Change Mitigation and Adaptation
    - Climate Change Mitigation
        - Trees absorb carbon dioxide from the atmosphere during photosynthesis, storing it in their wood and locking it away, which helps to reduce greenhouse gas levels substantially
    - Climate Change Adaptation
        - Forests can also help ecosystems and communities adapt to climate change by,
        - Enhancing Ecosystem Services: Preserving water quality, stabilizing air temperature, and improving water infiltration by absorbing moisture
        - Increasing Biodiversity: Creating suitable habitats and facilitating the migration of species by increasing landscape connectivity

        - Providing Safety Nets: Offering food and other products to local communities during times of crop failure due to climate change

- Insignificant Controversies and Challenges
    - Albedo Effect
        - In high-latitude Northern regions, planting trees can darken the landscape, reducing the amount of solar radiation reflected into space (albedo effect), which can offset some of the cooling effect and negatively impact the climate
    - Biodiversity Impacts
        - Large-scale afforestation can threaten existing ecosystems, especially low-growth ecosystems like grasslands, and can disrupt habitats for ground-nesting birds, as seen in Iceland
    - Land Use Competition
        - Planting forests can compete with other land uses, including food production, leading to potential land scarcity and conflicts

16.12.3 Schools of Thought on Biodiesel

- Biodiesel plantations can mitigate climate change by absorbing carbon dioxide during feedstock growth, which offsets some emissions from combustion. However, this benefit is undermined by controversies surrounding deforestation, food security, and inefficient land use
- Role of biodiesel plantations in climate change mitigation and adaptation
- Mitigation
    - Reduced emissions: Biodiesel can reduce greenhouse gas (GHG) emissions by 50–85% compared to fossil diesel over its life cycle. This is because the carbon dioxide released when the fuel is burned is part of the biogenic carbon cycle, which is absorbed during plant growth

  - Decreased fossil fuel reliance: The production and use of biodiesel decrease dependence on finite fossil fuels
  - Waste-to-energy: Some biodiesel is made from waste products like used cooking oil, which prevents this waste from releasing methane in landfills
- Adaptation
  - Diversification of fuel sources: Biodiesel offers a domestic fuel source, which can increase a nation's energy security and resilience to global market shocks
  - Climate-resilient feedstocks: Research is exploring feedstocks like algae and perennial grasses that can be grown on degraded or marginal lands, potentially with less water, to adapt to climate change impacts
- Controversies and negative impacts
  - Land use change and deforestation: Clearing forests or grasslands for plantations releases large amounts of stored carbon into the atmosphere, which can negate any climate benefits for decades. This is particularly problematic with crops like palm oil in tropical regions
  - Food security risks: Using food crops (corn, soyabean) for fuel production can drive up food prices and increase global food insecurity. This redirects resources from feeding people to powering vehicles

16.13 Terminology on Climate Change Studies

- Anthropogenic: A term referring to changes or effects resulting from human activities. Example: "Anthropogenic emissions" are greenhouse gases released by humans
- Blue economy concept: The concept seeks to promote economic development, social inclusion, and the preservation or improvement of livelihoods while at the same time ensuring environmental sustainability of the oceans and coastal areas
- Carbon capture and storage: The process of trapping carbon emissions produced by fossil fuel power plants or other industrial processes before they can enter the atmosphere by storing them deep underground

- Carbon Dioxide Equivalent: Carbon dioxide equivalent (CO_2 e) is a unit of measurement used to express the impact of different greenhouse gases on global warming on a common scale. It allows for the comparison of various greenhouse gases, such as methane (CH_4) and nitrous oxide (N_2O), to the warming potential of carbon dioxide (CO_2). A specific amount of a different greenhouse gas is converted to the equivalent amount of CO_2 that would have the same warming effect over a given period, typically 100 years
- Carbon dioxide removal: Carbon dioxide removal methods refer to processes that remove carbon dioxide from the atmosphere by either increasing biological sinks of carbon dioxide or using chemical processes to directly bind carbon dioxide . CDR is classified as a special type of mitigation
- Climate finance: Refers to financial resources and instruments that are used to support action on climate change
- Carbon footprint: Measure of the exclusive total amount of emissions of carbon dioxide that is directly and indirectly caused by an activity or accumulated over the life stages of a product
- Climate governance: The structures, processes and actions through which private and public actors seek to mitigate and adapt to climate change
- Climate Justice: Means putting equity and human rights at the core of decision-making and action on climate change
- Climate overshoot: Refers to the period during which warming will have increased past 1.5° C, before falling back down. This period will probably occur around the middle of this century, but troubling signs are emerging that it may occur even earlier
- Conference of the Parties (COP): The annual international climate conference under the UNFCCC, bringing together countries that have signed the convention
- Carbon removal: The process of removing greenhouse gas emissions from the atmosphere, through natural solutions such as reforestation and soil management or technological solutions like direct air capture and enhanced mineralization

- Climate resilience: This is the capacity of a community or environment to anticipate and manage climate impacts, minimize their damage, and recover and transform as needed after the initial shock
- Climate resilient development: This refers to the process of implementing greenhouse gas mitigation and adaptation measures to support sustainable development for all
- Climate-resilient pathways: Iterative processes for managing change within complex systems to reduce disruptions and enhance opportunities associated with climate change
- Climate security: Refers to evaluating, managing, and reducing the risks to peace and stability brought on by the climate crisis
- Eutrophication: Over-enrichment of water by nutrients such as nitrogen and phosphorus. It is one of the leading causes of water quality impairment. The two most acute symptoms of eutrophication are hypoxia (oxygen depletion) and harmful algal blooms
- Fossil fuel: Fossil fuels are Carbon-based fuels from fossil hydrocarbon deposits, including coal, oil and natural gas
- Feedback loop: Climate feedback loops happen when one change in the climate triggers further changes, in a chain reaction that reinforces itself as time goes on. Ultimately, feedback loops can trigger tipping points, at which point the changes to our planet's climate systems become severe and irreversible
- Global Warming Potential (GWP): The measure of how much heat a GHG traps in the atmosphere over a specific period, relative to carbon dioxide
- Nationally Determined Contributions (NDCs) **:**Nationally Determined Contributions (NDCs) are climate pledges and action plans that each country is required to develop in line with the Paris Agreement goal of limiting global warming to 1.5° C. NDCs represent short to medium-term plans that are updated every five years with higher ambition on climate
- Radiative Forcing
    - o The change in the net, downward minus upward, radiative flux (expressed in W m−2) at the tropopause or top of

atmosphere due to a change in an external driver of climate change, such as a change in the concentration of carbon dioxide, the concentration of volcanic aerosols or the output of the Sun.

  - The traditional radiative forcing is computed with all tropospheric properties held fixed at their unperturbed values, and after allowing for stratospheric temperatures, if perturbed, to readjust to radiative-dynamical equilibrium.
  - Radiative forcing is called instantaneous if no change in stratospheric temperature is accounted for.
  - The radiative forcing once rapid adjustments are accounted for is termed the effective radiative forcing.
  - Radiative forcing is not to be confused with cloud radiative forcing, which describes an unrelated measure of the impact of clouds on the radiative flux at the top of the atmosphere

- Renewable energy: Energy derived from natural sources that are constantly being replenished, such as wind, sunlight, the flow of moving water, and geothermal heat
- Rewilding: The mass restoration of ecosystems that have been damaged by human activity. More than conservation, which focuses on saving specific species through dedicated human intervention, rewilding refers to setting aside large areas for the natural world to regenerate in on its own terms
- Tipping Point: A threshold after which certain changes caused by global warming and climate change become irreversible, even if future interventions are successful in driving down average global temperatures. These changes may lead to abrupt and dangerous impacts with very serious implications for the future of humanity and our planet

16.14 Addressing Climate Change Challenge on Global Issues

- The individuals, families, communities, governments, organizations etc., have the best sciences available to plan for and implement climate-smart and resilient practices. Two critically important approaches which are available to managing climate resilience are mitigation and adaptation. These strategies are inter-related and must be pursued simultaneously as both can potentially improve resilience to the changing climate.

Chapter 17

Agrometeorological Concepts for Climate Change and Agriculture

"In Vedic thought, responsible decision makers, Scientists through their actions and knowledge are tasked with preserving the divine balance of Earth (bhumi), protecting the life-giving Weather (vayu), and ensuring sustenance through secure food systems (anna), recognizing the interconnectedness of all life and upholding Dharma for the well-being of all creatures"

17.1 Introduction to Importance of Concepts on Climate Change and Agriculture

- Developing scientific concepts to understand the negative impacts of climate change is crucial for several interconnected reasons
- Guiding mitigation and adaptation strategies
  - o Scientific concepts help in identifying and defining the specific risks and challenges posed by climate change, such as rising global temperatures, altered weather patterns, sea level rise, droughts, floods, and their effects on crops and livestock production, humans, and agroecosystems
  - o Scientific understanding of climate change impacts empowers informed decisions for mitigation (reducing emissions) and adaptation (adjusting to the changes)

- Scientific concepts and technological innovations are essential for creating solutions like sustainable agriculture and strategies for carbon capture
- By understanding which regions and populations are most vulnerable, resources and adaptation efforts can be prioritized to protect those at risk

- Raising awareness and promoting action
  - Scientific concepts help explain the causes and consequences of climate change to all the stakeholders
  - By providing evidence-based information, scientific concepts help to address misinformation and skepticism about climate change
  - Scientific concepts are essential for guiding policymakers in developing effective climate policies (PCCC and MCCC) and addressing the social and political dimensions of climate change
- Advancing research and innovation
  - Concepts help to improve climate models and projections, leading to more accurate predictions of future impacts
  - Investing in scientific research fosters innovation in areas like use of micrometeorological principles
  - To know the needs of further research to better understand the negative impacts of climate change, especially at local levels and to develop effective responses
  - Concepts also help in addressing climate change in collaboration and integration of knowledge across various disciplines like agronomy, entomology, plant pathology, crop physiology, genetics, agricultural physics, soil sciences, agricultural chemistry, crop physiology, economics, social sciences, etc.
- Developing strong concepts with scientific understanding of the negative impacts of climate change provides the knowledge, tools, and strategies needed to lessen its effects, adjust to unavoidable changes, and create a more sustainable future

17.2 The Concept of Micro, Meso, and Macro Meteorology for Climate Change Studies

- Agricultural Meteorology
  - Definition: The study of those aspects of meteorology that have direct relevance to agriculture. The abbreviated form of agricultural meteorology is agrometeorology
- Micrometeorology
  - Definition: The science concerned with detailed examination on the microscale basis of the physical and meteorological factors that are taking place within the ecosphere. Ecosphere: The space just above the crop/ tree canopy and just below the root zone
- Environment
  - Definition: A collective form embracing all the conditions in which organisms live
- Sub-divisions of the environment
  - Physical or abiotic or non-living environment. Example: Drought
  - Biotic or living environment. Examples: Predators, parasites, soil organisms, insects, pests, etc.
  - Both environments affect the distribution of organisms in different habitats
- The environment is also divided into
  - Microenvironment: This is the immediate surrounding of the organism
  - Macroenvironment: This is the total of abiotic and biotic environments that surround the organism and its environment

17.2.1 Different Groups of Climates

- For easy understanding of the study of climate, it is expressed in three different ways. These are
  - Microclimate: The climate of relatively smaller areas. Examples: a cropped field, a classroom, etc.

- Mesoclimate: The climate of a location. This climate is intermediate between micro and macro climate. Examples: a valley, forest, etc.
- Macroclimate: The climate of national or global in nature. Examples: arid, semiarid, humid regions etc., across nations or global scales, etc.

17.3 Scale of observations for Agricultural Meteorological Purposes

- Typical characteristic distances of climatic scales on horizontal scale for agricultural meteorological studies are
  - Microscale: Less than 100m
  - Toposcale: Local scale 100 m - 3 kms
  - Mesoscale: 3 kms to 100 kms
  - Macroscale: > 100 kms
- For agricultural meteorological observations, the above scales are globally followed

17.4 Policy Concept on Climate Change (PCCC)

- "Annadata Sukhi Bhava"- "May the provider of food be happy"
- "Krishihi Dharma"- "Agriculture is righteousness"
- "Kshetrapati Devata"- "Landowner is Lord of the cultivated soil"

17.4.1 The Genesis

- Climate Change is happening
- It is negatively impacting agriculture and livestock
- There's no sugarcoating it
- As envisaged by the IPCC keeping warming to 1.5°C by 2040 AD will be hard. Really hard
- At the same time the IPCC reports make it clear that the world has scientific understanding, technological capacity, and the financial means to tackle climate change

- Therefore, what the world needs is the will of all stakeholders including and policy and law makers at different levels to participate in the unprecedented and concerted actions necessary to stabilize temperature rise below 1.5 degrees Centigrade to ensure profitable agriculture, safe food production, and achieving food security
- Also, there are substantial economic and development benefits from bold climate actions
- Even more important is limiting global warming to 1.5° is imperative
- Falling short would lock in climate impacts
- Catastrophic consequences the world would face are enormous
- Farmers have the clarity they need
- Now it's time for scientists, policy makers, decision makers, policy makers, and law makers to step up to the challenge
- Agriculture is the backbone, livelihood, largest employer, and economies of most of the countries of world
- The carbon dioxide is reduced to carbohydrates in the plant physiological process "Photosynthesis" and food is produced
- In this process solar radiation, water, and carbon dioxide play the main roles
- Scientifically, solar radiation, water, and carbon dioxide are non- monetary inputs in agriculture as these are available freely in nature
- All weather elements are also renewable sources of energy and integral parts of agriculture, more so the solar energy captured and stored by crops through photosynthesis is also renewable
- However, all these non-monetary weather (elements) inputs are negatively impacted by climate change and crop production is affected and food security is at stake
- With rapid advances in information and communication technologies, and their expansion into rural areas, the demand for authenticated, scientific, and uninterrupted dissemination of information on agroadvisories, daily weather, timely and

accurate weather forecasts, early warnings, weather and climate services and products etc., has significantly increased among farmers in the backdrop of climate change

- In addition, changed weather conditions are influencing the occurrence of new pests and diseases on crops and animals further impacting agriculture
- Also, climate change driven weather changes are playing significant role in influencing market prices particularly for perishable commodities
- The cost of cultivation of crops and animal rearing is tremendously increasing beyond the control and affordability of the farmers and returns are meagre, forcing the farmers to take extreme steps
- While weather/ climate cannot be controlled and modified at meso and macro scales, micrometeorological technics that modify weather can be adapted at small scales (individual farmer's field level) based on weather forecast information for profitable agriculture, as daily weather information and weather forecasts play vital roles in planning daily agricultural operations
- Scientific works and studies project that climate resilient crop production, i.e., by monitoring daily weather conditions; by following weather and climate information, services and products; and adopting micrometeorological technics during the growing season it is possible to reduce cost of cultivation of crops and animal rearing by 50 per cent; increase in yields by 50 per cent; and 10 per cent in quality improvement of crops and livestock
- By linking PCCC with MCCC it is very much possible to achieve these benefits and farmers income doubles under climate change scenarios

17.4.2 The Hypothesis

- Humans need food for survival
- Food is being produced by the farmers through their agriculture
- Agriculture is weather and climate dependent

- However, weather and climate are negatively impacted by climate change threatening agriculture, food production, and food security
- In addition, weather and climate are renewable energy resources and non-monetary inputs in agriculture
- Therefore, for profitable agriculture, safe and enough food production, and food security under climate change it is essential to provide services, products, and information on daily weather, early warnings, agromet advisories, climate resilient agricultural practices etc., to the farmers
- At present, the IMD, SAUs, ICAR Centers are generating and disseminating the services
- In future, Artificial Intelligence (AI) and Machine Learning (ML)/ Deep Learning models can produce localized forecasts for specific geographical areas
- In addition, hyper local weather forecasting at micro level is also possible with AI and ML
- At present few establishments have already started these weather forecasts
- To provide authenticated, uninterrupted, qualitative, and reliable information on these services to the farmers an Act and Statutory based "Link" between "Farmers" and "SAUs to provide information, services, and products on weather and climate" is needed
- At present this "Link "is missing
- It is possible for SAUs to provide this "Link" only when additional "Act and Statutes" are embedded into their existing Act and Statutes
- This needs enactment of concerned Law making Statutory and Legislative Bodies in respective States
- All the international Conventions, Treaties, Summits, etc., on climate change and agriculture emphasized to "Integrate climate change measures into national development programs, policies, strategies, and planning to protect climate systems from negative impacts of climate change"

- India is implementing several programs as specified in these Conventions, Treaties, Summits, etc.
- Therefore, it is time for the SAUs to consider developing their own "Climate Policies" on lines of National and their State policies
- The climate policy of SAUs paves the way to incorporate additional Act and subsequent Statutes into their existing Act and Statutes to provide the missing "Link" between "Farmers" and "SAUs to provide information, services, and products on weather and climate"
- In the climate policy of SAUs and in the additional Act and Statutes, provision be made available to incorporate a Clause on "To continue the ongoing Agromet advisories (AASs and AABs) with value addition and to undertake MOUs with international and national organizations working on weather and climate services for value addition to the ongoing agromet advisories and trainings to all stake holders of climate policy of the SAU for enhanced services to the farmers"
- If it is done and the SAUs adopt micrometeorological technics in synergy with ongoing agronomic, genetic, etc., strategies for climate change as emphasized in the MCCC the cost of cultivation of crops and livestock rearing can be reduced by 50 per cent in addition to an increase in the crop and livestock yields by 50 per cent, and also 10 per cent increase in quality of produce under climate change scenarios
- It is also essential to mention in the climate policy itself that additional and sustained funds be provided by appropriate authorities for its successful implementation
- Finally, this leads to profitable agriculture, continued and sustained food production, and food security under climate change scenarios and double the income of farmers as well
- The farmers who do their sacred duty (*dharma*) that sustains the world shall not be denied these services

17.4.3 Excerpts from International Conventions, Agreements, and Summits

- According to UNFCCC, *Madam Patricia Espinosa, the Executive Secretary of the UNFCCC in her discourse mentioned* "Integrate climate change measures into national policies, strategies, and planning"
- In addition, the UNFCCC underlines that climate change policies to protect climate systems should be "Integrated with national development programs" by taking all sources and sinks of carbon into consideration
- Compared with the UNFCCC, the Kyoto Protocol which is also linked to UNFCCC, focuses much more explicitly on agriculture and forestry on the "Promotion of sustainable forms of agriculture in the light of climate change considerations and on sustainable forest management practices"
- There are also commitments in the Protocol for countries to "Formulate national and regional programs" to improve scientific cooperation, education and training, inventories of emissions and sinks, and adaptation potential"
- The Paris Agreement which is a "Legally binding international treaty on climate change" stands as a cornerstone of international climate action, establishing a comprehensive framework for coordinated efforts
- India is committed to Paris Agreement as a signatory
- This legal bound Agreement calls all Parties to engage in "Adaptation, planning, and implementation through national adaptation plans"
- The Agreement is a global treaty which aims to restrict global warming level below 2 degrees centigrade above the pre-industrial levels and tries to limit it to 1.5 degrees Centigrade
- Targets under the Sustainable Development Goals (SDG) of climate action for Food production and Security also suggested "Integrate climate measures into national policies, strategies, and planning"

- The Convention on Biological Diversity (CBD), which is based on a "Broad ecosystem approach" contains three national level obligations viz., (a) conserve biological diversity (b) sustainably use biological diversity and (c) share its benefits
- The CBD recognized that there are major interactions between climate and biological diversity
- The United Nations Convention to Combat Desertification (UNCCD) suggested "Incorporation of long-term strategies to combat desertification and mitigate the effects of drought in the national action programs"
- The World Food Summit (WFS) emphasized and suggested "To maintain, promote, and establish preparedness strategies and mechanisms like development and application of climate forecast information for surveillance and early warning of droughts, floods, and pest and disease alertness"
- **Inference:** As emphasized and suggested in the global Conventions, Treaties, Summits, etc., on climate change and agriculture the SAUs may develop their own "Climate Policy" on lines of National and their State policies for sustainable agriculture, food production, food security, and above all in the interest of the farmers, the unselfish food creators and providers for all. The policy of SAUs paves way to incorporate additional Act and Statutes into their existing Act and Statutes to provide the missing "Link" between "Farmers" and "SAUs to provide information, services, and products on weather and climate". It is essential to mention in the policy itself that additional funds be provided and continued by appropriate authorities for its successful implementation

17.4.4 India's Proactive Approach

- India is one of the proactive countries in UNFCCC in terms of global climate change negotiations and implementing different climate actions at the ground level
- India had announced in Glasgow in November 2021 at UNFCCC's COP26 that by the year 2070, India will achieve the target of Net Zero Emissions and carbon neutrality

- With reference to agriculture, in ICAR- NAARM Policy Paper number 13 entitled "Agriculture Sector Contributions in Achieving Net Zero Emission by 2070" it is mentioned that "Effective coordination between the National Action Plan on Climate Change (NAPCC) and State Action Plans on Climate Change (SAPCC) is necessary for India to meet its promise to achieve Net Zero GHG Emissions from the agriculture sector by 2070" and suggested for NAPCC and SAPCC synergy
- In India the agriculture sector comes under State subject
- The ICAR- NAARM policy paper also opined that "Coherent and strong Centre- State collaboration plays an important role in achieving the emission reduction strategies in agriculture sector"

17.4.5 The Agromet Advisory Services in India

- Since ancient times, Indian economy is largely dependent on agriculture and weather plays major role in crop growth and yield
- The need for a close collaboration between the Departments of Agriculture and Meteorology in India was felt by the Royal Commission on Agriculture during 1920s
- This collaboration is to better understand the relationship between weather and crops, so that disruption and damage caused to agriculture, economy, and human life by weather hazards could be reduced, marginalized or even controlled from droughts turning into famine
- Agromet decision support system and the recent advances and major achievements
    - A dedicated portal "Agromet- DSS" i.e., http:/ agromet.ind.gov.in is developed during the recent years for exchange of seamless data and information among concerned organizations in a more objective, timely, transparent, and effective manner leading to improvement in the quality of services
    - Agromet – DSS facilitates the scientists and institutions with customized tools to integrate weather and crop information to prepare agromet advisories at finer scales in short time to serve the farming community at micro level

- This portal is presently used for value addition to district level weather forecasts by RMCs, and MCs as well as for preparation of district and block level AAS bulletins
- This portal is also used for updating weather information and agromet advisories in mobile app "Meghdoot"

- Future efforts
  - Demand for weather forecast at finer spatial scale (from block level to panchayat level) with longer duration (from 5 days to 10 days) started gaining momentum
  - To fulfil this requirement of the farming community, emphasis is being given on introduction of high-resolution models with appropriate accuracy along with assimilation of satellite data
  - Efforts are being made to use more data acquired from remote sensing sources, couple with weather information, for identification of crop stresses on a near real time basis
  - Further integrated approach is planned to involve all the leading organizations of India like IMD, NCMRWF, IITM, ICAR, ISRO, IITs, NIC, SAUs, SDAs etc.
  - It is also felt essential to take benefit from the fast-growing technologies to generate meaningful agromet advisories for next 10 days for the farming community at panchayat level
- **Inference:** In the climate policy of SAUs and in additional Act and Statutes to be embedded into the existing Act and Statues, provision may be made available to include a Clause as "To continue the ongoing Agromet advisories (AASs and AABs) with value addition".

17.4.6 Global and Indian Weather Forecasting Services

- Global forecasting services
  - World Meteorological Organization (WMO), Global. This is the UN agency coordinating global weather, climate and weather issues. Role: Sets international standards for weather data and forecasting, supports Global Framework for Climate Services (GFCS). Runs World Weather Watch and Service Weather Forecasting Program (SWFP)

- National Weather Service (NWS), USA. Services: Daily and severe weather forecasts, hurricane warnings (via National Hurricane Centre); climate prediction (via Climate Prediction Centre). Plat forms: Weather.gov, mobile alerts, NOAA weather radio
- European Centre for Medium Range Weather forecasts (ECMWF), Europe. Intergovernmental organization supported by 35 + countries. Services: Global medium to long range forecasts, seasonal climate forecasts, reanalysis datasets (ERA5). Platform: ECMWF provides data feeds to national agencies across Europe and beyond
- UK Met Office, United Kingdom. The UK's national meteorological service. Services: Weather forecasts (local to global scale), climate monitoring, aviation and marine services, defense and energy sector forecasts. Platforms: Met Office web site, mobile app, TV bulletins
- Meteo France- France. National meteorological service of France. Services: Public weather forecasting, marine and mountain weather alerts, specialized aviation forecasts, climate research and modelling
- Japan Meteorological Agency (JMA)- Japan. National weather agency of Japan. Services: Weather and tsunami warnings, earthquake monitoring, typhoon tracking, and seasonal climate forecasts
- Bureau of Meteorology (BoM) – Australia. Australia's national weather, climate, and water agency. Services: Weather warnings and daily forecasts, cyclone alerts, seasonal outlooks (rainfall, temperature), climate change monitoring
- Environment and Climate Change Canada (ECCC)- Canada. Services: Weather forecasts and warnings, severe weather alerts (floods, snowstorms), marine and aviation weather services, climate change science and data
- Deutscher Wetterdienst (DWD) – Germany. National meteorological service of Germany. Services: Public

weather forecasts, severe weather warnings, aviation services, climate observation and advisory

- Private weather companies (global). Examples: The Weather Company (IBM), Accu weather, Tomorrow.io Services: Real time global weather forecasts, hyperlocal forecasting (location specific), customized weather services for businesses (agriculture, energy, logistics)

17.4.7 Indian Weather Forecasting Services

- India Meteorological Department (IMD). Services: Daily weather forecasts (short, medium and long range); cyclone tracking and warnings, aviation and marine forecasts, agrometeorological services (Agromet), flood forecasting support. Platforms: Mausam app, Meghdoot App (for farmers), Damani App (for lightening alerts)
- Indian Space Research Organization (ISRO). Services: Supports weather monitoring using satellites like INSAT and METSAT (Kalpana-1), provides satellite imagery, data on rainfall, cloud cover, sea surface temperatures. Works closely with IMD for better forecasting accuracy
- National Center for Medium Range Weather Forecasting (NCMRWF). Under Ministry of Earth Sciences (MoES). Services: Medium range (3-10 days) weather forecasts, agro-advisory services, model-based forecasting (using high performance computing)
- Indian Institute of Tropical Meteorology (IITM). Pune based institute of MoES. Services: Monsoon research, climate modelling, seasonal forecasts (El Nino/ La Nina predictions), runs programs like monsoon mission for improving weather models
- State Agricultural Universities (SAUs) and Krishi Vignan Kendras (KVKs). These collaborate with IMD under Agromet Advisory Services Program, provide local level weather forecasts and crop- specific advisories

- o Research Institutions and Universities. IITs (Delhi, Pune, Bhuvaneswar) and other institutions also develop new models, conduct climate studies, and improve forecast technology
- o Central Water Commission. Services: Provides flood forecasts. Monitors river water levels and rainfall to issue early warnings
- o Disaster Management Authorities (NDMA, SDMAs). Services: Use forecasts of IMD and other agencies to issue public disaster warnings, coordinate during extreme weather events like floods, cyclones, and heat and cold waves
- o Private weather companies. Examples; Skymet weather, Weather Risk Management Services (WRMS), Earth Networks India. Services: Customized forecasts for agriculture, insurance, energy sectors; mobile Apps and SMS alert services, risk management consultancy based on weather predictions

- **Inference:** In the climate policy of SAUs and in additional Act and Statutes to be embedded into the existing Act and Statues, provision may be made available to include a Clause as "MOUs may be undertaken with international and national organizations involved in weather and climate services for value addition to the ongoing agromet advisories and to provide trainings to all stake holders of climate policy of the SAUs for enhanced services to the farmers"

17.4.8 Process and Procedure of Decision Making on "Climate Policy" at SAUs Level

- Core points for consideration in decision making on climate policy development and implementation
  - o The core points and main text for consideration in decision making and for inclusion in the "Climate policy" and for enacting an additional Act and subsequent Statutes into the existing Act and Statutes of SAUs are,
    - ▪ The farmers who perform their sacred duty(*dharma*) of creating food that sustains the world shall not be denied providing weather and climate information, services and products, agromet advisories, daily weather, early warnings, climate resilient agricultural practices etc., under the ongoing climate change scenarios

- As emphasized and suggested in the global Conventions, Treaties, Summits, etc., on climate change and agriculture the SAUs may develop their own "Climate Policy" on lines of National and their State policies for sustainable agriculture, food production, food security, and above all in the interest of the farmers, the unselfish creators and providers of food for all. The policy of SAUs paves way to incorporate additional Act and Statutes into their existing Act and Statutes to provide the missing "Link" between "Farmers" and "SAUs to provide information, services, and products on weather and climate". It is essential to mention in the policy itself that additional funds be provided and continued by appropriate authorities for its successful implementation
- In the climate policy of SAUs and in additional Act and Statutes to be embedded into the existing Act and Statues, provision may be made available to include a Clause as "To continue the ongoing Agromet advisories (AASs and AABs) with value addition". These advisories proved phenomenal success in the service of farmers
- In the climate policy of SAUs and in additional Act and Statutes to be embedded into the existing Act and Statues, provision may be made available to include a Clause as "MOUs may be undertaken with international and national organizations involved in weather and climate services for value addition to the ongoing agromet advisories and to provide trainings to all stakeholders of climate policy of the SAUs for enhanced services to the farmers", the need of the hour to protect the noble profession of the farmers under climate change

- Other points and issues for consideration in climate policy development and implementation

- Policy development
- In addition to the points mentioned under "Process and procedure of decision making on climate policy at SAUs level" as above, the following issues be considered
  - Challenges of India's vis-à-vis State agriculture sector's range from climate change and variability, nutrition security, and transboundary crop and livestock to diseases, etc., which demand convergence
  - Therefore, the climate policy of SAUs to deal with climate change receive top priority equally by SAUs and respective Governments because climate change affects agriculture, allied sectors, strategic food production, and food security
  - Focused policy attention to agriculture and allied sectors like horticulture, livestock, fisheries, etc., be in place
  - For climate policy "ONLY", integrating crops, livestock, and fisheries under one umbrella is a top priority as veterinary and fisheries are integrally linked with agriculture (this integration may not be beyond climate policy)
  - The climate policy be envisaged under a single institutional framework, which brings together diverse domines like agriculture, horticulture, veterinary, etc., in SAUs, Departments of Agriculture and allied sectors in the State, and IMD, SDA, NDA, etc. (this convergence may be only for climate policy)
  - Provision be made in the policy for effective coordination among the National Action Plans on Climate Change (NAPCC), State Action Plans on Climate Change (SAPCC), and the climate policy of the SAUs
  - The climate policy shall focus on capacity building to all stakeholders viz., farmers, scientists, teachers, extension specialists, policy makers, administrators, law makers, etc., at local, national, regional, and international levels
  - The need to incorporate programs to improve scientific co-operation, education, and training through

interdisciplinary approach among all stakeholders shall be examined

- Similarly, need for sharper coordination, stronger leadership, and better program design are in place
- Also, the need for consolidation and integrated approaches among all the disciplines in climate policy shall be emphasized as National Education Policy (NEP) 2020 of GOI calls for integrated multidisciplinary institutions and discourages narrow and compartmentalized structures, simultaneously emphasizing collaboration across domines
- The climate policy shall provide interdisciplinary focus ensuring that cross-cutting challenges on climate change are addressed comprehensively
- There is a need to make provision for administrative clarity and long-term systemic coherence
- Guidelines be developed for robust monitoring and evaluation which are essential to track progress and identifying the areas where corrections are needed
- A dedicated climate policy be seen to boost SAUs visibility, attract tailored funding, and improve responsiveness
- Therefore, provisions be made available for SAUs to submit proposals to State, National and international organizations, institutions, etc. for funds
- The climate policy be institutionally embedded across all disciplines, faculties, etc., in the SAUs
- The climate policy of the SAUs be seen as "Internal integration"

- Policy implementation
  - The challenge lies in translating effective policies and strategies into action by the SAUs
  - It is essential that the best possible Act and Statutory bound daily weather information, services, products, early warnings, forecast services, agromet advisories, climate resilient agricultural practices, etc., be prepared by the SAUs under the guidance of a group of experts and systematically and uninterruptedly delivered to the farmers

    - It is difficult to start action with the existing system and facilities
    - Therefore, appropriate authorities consider providing additional and enough infrastructure and sustained finances to the SAUs
    - Even to start on a pilot basis, additional and sustained finances are essential which may be provided by appropriate authorities
    - It is also essential that additional agrometeorlogists are recruited on permanent/ regular basis for preparation and dissemination of daily weather information, services, products, early warnings, forecast services, agromet advisories, climate resilient agricultural practices, etc.
- The Making of Additional Act and Statutes
    - Bottom to top Approach: The SAUs may or as appropriate,
        - Organize and take decisions through meetings at the University level as per the existing Act and Statutes. Example: Faculty Board, Academic Council, relevant research, extension, administrative, etc., bodies
        - Take the approval of the Board of Management of the University
        - Submit the proposals to the respective authorities for enactment, as appropriate
        - The additional Act be embedded into the existing Act and Statutes, and additional Statutes be developed by SAUs
        - Involve the finance committee of the university at all levels
        - Implement the additional Act and new Statutes by the SAUs
    - Top to Bottom Approach: Same procedure in the reverse manner, or as appropriate
- Through the above initiatives, processes and procedures, it is possible to provide the missing "Link" between "Farmers" and "SAUs to provide information, services, and products on weather and climate". By this way, it is possible to reduce cost of cultivation of crops and animal rearing by 50 per cent;

increase in yields by 50 per cent; and 10 per cent in quality improvement of crops and livestock produce, and to double the income of farmers, under climate change scenarios in synergy with MCCC

17.5 Micromet Concept for Climate Change (MCCC)

- *"Krishim vina na jivanti jivah sarve pranashyati"*- "Without agriculture, no living being survives"
- *"Shunam krishathu langalam"*- "Let the ploughshare furrow the earth comfortably"
- "May the lands so plowed and rich with milk and wet with honey and ghee come back to us much fertilized"
- "The farmer is the one who communes daily with Earth, sky, and seed"
- *"The farmer is also the one who carries the burden of the king on his shoulders and provides for the sustenance of all others"*

17.5.1 The Genesis

- In "Soil, plant, and weather continuum (SPWC)" it is essential that perfect "Soil health, plant health, and weather health" are at optimum levels for enhanced crop yields
- According to IPCC,
  - o The soil health, plant health, and weather health (atmosphere) are alarmingly affected by global GHG emissions which are injected into the atmosphere resulting in climate change
  - o Global emissions are projected to be 52-58 Gt CO_2e by 2030
  - o Annual emissions need to be about half that (25-30 Gt CO_2e/yr on average) by 2030 to limit warming to 1.5°C (with no or low overshoot)
  - o There is a need to focus efforts not only on reducing emissions, but also removing and storing carbon from the atmosphere for profitable agriculture

  - Carbon removal is necessary for both moving to net-zero emissions and producing net-negative emissions to compensate for any overshoot of 1.5°C
  - If global warming is not limited to 1.5 degrees centigrade in the nearest future, by 2040, the agricultural production systems get affected threatening food security
  - Simultaneously, as stated above, it is essential that soil health, plant health, and weather health be achieved for continued and profitable agriculture in the backdrop of changing climate
  - The IPCC also reported that while it is still technically feasible to avoid a 1.5°C rise in temperature, ***"Technologies will need to shift across the board"*** to achieve these emissions reductions for sustained agriculture
- This observation of IPCC is the basis and genesis of Micromet Concept for Climate Change (MCCC)

17.5.2 The Hypothesis

- The IPCC reported that global warming due to climate change is happening, and it is unequivocal
- Climate change is negatively impacting all weather elements
- Weather elements are essential for photosynthesis
- Photosynthesis is adversely affected by abrupt changes in weather elements due to climate change
- This in turn is resulting in reduced crop yields
- This is not desirable for profitable agriculture, food production, and food security
- Technics and technologies available on date indicate that it is not possible to modify/change the weather/ climate on large scales (meso and macro scales) and larger areas in agroecosystems to suit agricultural production under climate change
- Under such situations, adjusting agricultural operations depending on prevailing weather conditions is one of the options

- This is just one of the several tactical options which is not scientific, only a temporary option that partially addresses the problems
- The Vedic science enlighten that exchange of energies, interrelationships and interactions exist among the "Panch Bhutas" i.e., "Natural Forces" viz., Akasha/Aether, Vayu, Agni, Apah/ Jal, and Prithvi
- Among these Akash/Aether, Vayu, Agni, Apah/ Jal influence crops both on aerial and underground parts of crops with "Prithvi" as base
- By understanding this exchange of energy, interrelationships, and interactions ancient farmers were successful in their agriculture
- They were also able to protect their agricultural and livestock crop production bases, systems, environment, and natural resources from degradation
- The modern agrometeorologists also recognized the exchange of energy, interrelationships, and interactions among the same natural weather elements (Panch Bhutas/Natural Forces) viz., Solar radiation (Akasha/Aether), Air (Vayu), Temperature (Agni), Rainfall (Apah/Jal) and Cultivated soil (Prithvi)
- The modern agrometeorologists also proved that the exchange of energy, interrelationships, and interactions occur both in the aerial and underground parts of crops with cultivated soil as base
- It is also proved that a variation in one or a combination of more than one weather element that negatively affects photosynthesis, growth, and development of crops under climate change cannot be modified in the agroecosystems at meso and macro levels
- However, both Vedic science and modern agrometeorologists proved that it is very much possible to modify these weather elements at micro level (farmer's field level) by following micrometeorological principles
- In addition, way back in 1920s the Royal Commission on Agriculture (RCA) recognized the importance of research on micrometeorological studies of crops and suggested for its implementation to overcome acute food shortages

- New and impactful micrometeorological technics and technologies can be developed taking into consideration the transfer/exchange of energy using modern physical principles as basis
- Therefore, these may be developed by modern agrometeorologists for continued success of agriculture under climate change
- At the same time there is a need for synergy between "Technics and technologies being developed by agronomists, soil scientists, plant breeders, etc." and "The new micrometeorological technics and technologies of modern agrometeorologists" in agricultural sciences to effectively combat the negative impacts of climate change on weather elements, photosynthesis, and crop yields
- For enhanced and effectiveness of this process there exist a need to develop and implement a "Climate Policy" of SAUs that include and read as "Micromet Concept for Climate Change (MCCC)" on the lines of PCCC
- An additional Act and subsequent Statutes that read" Microclimate Concept for Climate Change" may be incorporated into the existing Act and Statutes of the SAUs
- This makes provision for issuing executive orders to the concerned by SAUs under the "Subject" as "Micrometeorological technics and technologies as appropriate for climate change be continuously developed and provided to all stakeholders of agriculture on priority to counter the adverse impacts of climate change"
- Finally, the synergy of both PCCC and MCCC effectively results in profitable agriculture, assured food production, and food security under climate change
- In addition, by implementing these concepts, simultaneously and synergy, it is possible to reduce the cost of cultivation of crops and livestock rearing by 50 per cent, in addition to an increase in the crop and livestock yields by 50 per cent, and 10 per cent increase in quality of produce under climate change scenarios
- Additionally, the simultaneous and synergy impact of PCCC and MCCC leads to double the income of farmers as well

17.5.3 Excerpts on Climate Change, Weather Elements, Cultivated Soils, and Photosynthesis

- Climate change is negatively impacting all weather elements
  - Solar Radiation (Akasha/Aether)
    - Reduction in solar radiation at 2 per cent per decade at many stations in India during the period of May to December is reported
    - It is also reported that during the past 50 years in many industrial regions of the globe solar radiation has decreased by 2.7 per cent per decade
    - This is due to an increase in cloudiness because of anthropogenic aerosols and pollutants changing the optical properties of the atmosphere
  - Wind (Vayu)
    - Climate change is intensifying some extreme weather events including wind gusts associated with thunderstorms and hurricanes
    - Climate change could affect wind, with some regions experiencing decreases in wind speeds and others potentially seeing increases, depending on the specific location
  - Temperature (Agni)
    - The temperature in the first two decades of the 21st century (2001-2020) was 0.99 °C higher than 1850-1900
    - Global surface temperature was 1.09 °C higher in 2011–2020 than 1850–1900
    - The temperature increase will cross the 1.5degree centigrade threshold by 2040, which was projected much later
  - Rainfall (Apah/Jal)
    - Heavy precipitation events have become more frequent over most areas during the twentieth century due to climate change
    - The areas affected by drought have increased in many regions since 1970
    - Intense tropical cyclone activity has increased in some regions since 1970

    - Observations show that changes are occurring in the amount, intensity, frequency, and type of precipitation/ rainfall
- Cultivated Soil (Prithvi)
    - The dynamic exchange of energy, momentum, heat, and matter/mass between the land surface and the weather includes interaction among "Solar Radiation(photons), Wind (momentum), Temperature (thermal energy /heat), and Water(matter/mass)"
    - In addition, interactions among the near-surface weather, agricultural crops, and the soil environment at micro level are crucial for understanding and optimizing agricultural practices which lead to improved crop yields and sustainable land management
    - Also, micrometeorological conditions affect all crop development stages viz., germination, flowering, seed setting, and final crop yield
    - Turbulent transport and energy exchange among the weather, the biologically active canopy and upper soil surface takes place
    - In summary, the micrometeorology of cultivated soils highlights the vital connections among weather, plants, and soil in agricultural systems. By understanding and leveraging these interactions, it is possible to improve crop production, optimize resource use, and promote more sustainable farming practices at micro level under climate change scenarios
- Photosynthetic process
    - All the weather elements play significant role in the process of photosynthesis
    - Solar radiation controls life by heating the earth and atmosphere, and by providing the energy required in photosynthesis for conversion of carbon dioxide and water into the primary sources of food (carbohydrates)
    - In the process of photosynthesis, carbon dioxide and water are consumed, and organic matter (sugar) and oxygen are formed, and radiant energy is stored as potential energy in the oxidizable organic molecules

- The oxygen liberated comes from the water consumed
- Thus, in this process the stable water molecule is cleared (photolyzed) by radiant energy absorbed in the chloroplast
- The resulting electrons are passed through intermediary carriers and used to reduce carbon dioxide to the level of carbohydrates
- Gas exchange in the leaf occurs through stomata while water is made available
- This clearly indicates that weather health is essential for photosynthesis

17.5.4 Vedic Science on "Pancha Bhootas/Five Natural Forces"

- During the Vedic period farmers believed that if they prayed to the forces of nature which are Gods that were unknown and unexplained (known only by divine grace) they are protected and their agriculture flourishes
- These most powerful forces are called "Pancha Bhootas/ Five Natural Forces" viz., Akasha/Aether (Solar radiation), Vayu (Air), Agni (Fire), Apah or Jal (Water), and Prithvi (Earth)
- Among all these, the farmers relied to the highest level on the blessings of "Earth" i.e., "Mother Earth" and worshipped as Goddesses
- The reason is that according to Vedas all the other four forces exist because of "Earth"
- In addition, the earth is the only most pious, sacred, and revered planet known to support "Life" so far in the Universe, which is worshipped also as "Mother Earth (Prithvi)"
- The "Sky" is the "Father" who is the chief of the deities, hence the name "Father Sky (Pitaa)"
- Vedic Science on the characteristics of the "Pancha Bhootas/ Five Natural Forces"
    - o Akasha (Aether): The sky has only one character viz., "Sound". In modern agrometeorology it is "Solar radiation dynamics" which also include thunder, lightning, etc. The

solar radiation dynamics are explained in classical agrometeorology through several solar radiation laws

- o Vayu (Air): The air has two characteristics viz., "Sound and Sense". In classical agrometeorology it is "Aerodynamics" and mostly explained by Newton's laws
- o Agni (Fire): The fire has three characteristics viz., "Sound, Sense, and Shape". In classical agrometeorology it is "Thermodynamics" and mostly explained by Fourier's law
- o Apah / Jal (Water): The water has four characteristics viz., "Sound, Sense, Shape, and Taste". In classical agrometeorology it is explained both by "Hydrological cycle" and by "Diffusion processes" of Fick's law
- o Prithvi (Earth): The soil has five characteristics viz, "Sound, Sense, Shape, Taste, and Smell". In classical agrometeorology soil is "Soul of infinite life" and it is proved that all the above laws are integral part of "Earth" and "Weather" interactions
- o All natural forces are interrelated and interact among themselves
- o Vedic science also enlighten that the exchange of energies, interrelationships, and interactions among Akasha/ Aether (Sky), Vayu (Air), Agni (Fire), Apah or Jal (Water), occur both in the aerial and underground parts of crops with Prithvi (Earth) as base

- Vedic Science on exchange of energy, interrelationships, and interactions among "Pancha Bhootas/ Five Natural Forces"
    - o The mother "Prithvi (Earth)" gave all her five characteristics to her children as well. Her children include those who all depend on her viz., crops, humans, livestock, trees, mountains, rocks, etc., and this is the reason that all these living and non-living things are shouldered and well protected in her hands by her. These children live comfortably, and life goes on and on with the care and protection of Mother Earth because of continuous exchange of energy, interrelationships, and interactions among five Natural forces. The classical agrometeorology explains this very elaborately
    - o The "Apah/Jal (Water)" sacrificed one of the five characteristics that Mother Earth possesses viz., "Smell" and

retained only four. That is why neither Mother Earth nor her children can hold water. The water escapes and cannot be retained in hands, soil or stored by the plants etc., and in the modern science the "Hydrological cycle" goes on and on. Fick's law adds fine touches in addition to several non-dimensional numbers explaining this extensively

- The "Agni (Fire)" sacrificed two of the characteristics that Mother Earth possesses viz., "Smell and Taste" and retained the other three. That is why no one can touch "Fire". In modern science this is "Heat / thermal energy" which the classical agrometeorology explains through the thermo-dynamics, Fourier's law, and other laws
- The "Vayu (Air)" sacrificed three of the characteristics that Mother Earth possesses viz., "Smell, Taste, and Shape" and retained the other two. That is why no one can see the "Air". It can only be felt. In classical agrometeorology it is explained by innumerable laws of aerodynamics including Newton's law of viscosity
- The "Akasha/Aether (Sky) sacrificed four of the characteristics that Mother Earth possesses viz., "Smell, Taste, Shape, and Sense" and retained the left over one i.e., "Sound". That is why the "Sky" is only a human imagination. In the classical agrometeorology the energy (digital) dynamics, sound dynamics etc., explain this, in addition to several laws of "Solar radiation dynamics". The Sun also continuously burns Himself to provide very sacred, most pious, and unfathomable energy for the children on the earth
- The Vedic sciences enlighten that all the five Natural Forces are interrelated and interact ceaselessly. In addition, "Akasha (Aether), The "Vayu (Air)", The "Agni (Fire)", and "Apah/ Jal (Water)" also exist in "Prithvi (Earth)"
- Therefore, "Prithvi (Earth)" has life and provide mechanical support, water, energy for gaseous exchange, nutrients etc., to the underground / aerial parts of the crops for successful crop production
- The "Prithvi (Earth)" also acts as base for exchange of energies above and below it among "Akasha/Aether (Sky), The "Vayu (Air)", The "Agni (Fire)", and "Apah /Jal (Water)" for growth and development of crops

- Examples of Micrometeorological technics of Vedic/Ancient civilizations
- Vedic/Ancient civilizations used various ingenious methods to modify their microclimates for agricultural crops, often based on a deep understanding of their local agroecosystems. These traditional micrometeorological techniques helped manage temperature, moisture, wind, and solar radiation to improve crop yields and sustainable agriculture
    - Microclimate modification through physical structures
        - Terracing: Most famously practiced by Asian farmers and the Inca in the Andes, terracing was also widely used in places like ancient China and the Mediterranean. The technic involves creating level platforms on steep hillsides, supported by stone walls. The microclimate effect: The stone walls of the terraces absorb solar radiation during the day and radiate heat at night, protecting crops from cold mountain air. Terraces minimize erosion and conserve soil and water by reducing runoff. The varying altitudes of the terraces create multiple microclimates, allowing the Asians, Mediterranean, and Inca to cultivate a wide variety of crops
        - Raised fields and chinampas: The Pre-Columbian civilizations, particularly the Aztecs in Central Mexico, constructed sophisticated floating gardens called *chinampas*. The technique involves building rectangular plots of land on shallow lake beds using layers of mud, vegetation, and decaying organic matter. Microclimate effect: The surrounding canals and water bodies regulate the temperature, protecting crops from frost by releasing heat absorbed during the day. The water provides constant moisture to the crops, reducing the need for manual irrigation

  - Stone mulch (Lithic mulch): The Asians and Nabataeans in the Negev Desert used stone mulch. The technique involves placing a layer of stones on the ground between plants. Microclimate effect: The stones act as a buffer against extreme desert temperatures, accumulating heat during the day and releasing it at night. They trap moisture by reducing evaporation and suppress weed growth
- Microclimate modification with water and soil
  - Irrigation and water harvesting: In Asia, Ancient Mesopotamia, Egypt, and Nabataea this was followed. The technique involves developing complex systems of canals, dikes, aqueducts, and rainwater harvesting cisterns. Microclimate effect: Irrigation directly add moisture to crops and the surrounding soil, increasing humidity and lowering local air temperatures. In arid regions like Nabataea, Western Asia, and parts of Central Asia cisterns and terraces were built to capture and absorb runoff from the infrequent rains
  - Organic soil amendments: The technique involves incorporating organic matter like compost, manure, and decomposed plants into the soil. Microclimate effect: This significantly increases the soil's water-holding capacity, allowing it to act like a sponge and maintain more consistent moisture levels. Organic matter also improves soil structure and reduces erosion
- Microclimate modification using plants
  - Shade and intercropping: This is common in many ancient civilizations, such as the South East Asia, Mayans, and the ancient Egyptians. The technique involves the concept of "Three sisters" wherein the Native Americans planted corn, beans, and squash together. The tall corn provides shade for the lower-growing and moisture-loving beans and squash. Milpa systems in Mesoamerica, where crops were interspersed with fruit trees provided shade and improved

microclimates for the food crops. Microclimate effect: Shade from taller plants reduces direct sun exposure, conserving moisture in the soil, and preventing heat stress in sensitive crops

- Windbreaks and shelterbelts: These are used in Ancient Asia, China, Egypt, and recorded in Scotland as early as the 15th century. The technique involves planting rows of trees and shrubs to create barriers against wind. Microclimate effect: Windbreaks reduce wind speed, which in turn reduces water loss from transpiration by crops and soil through evaporation. A protected microclimate is created with increased soil temperature and more evenly distributed moisture

- **Inference:** The Vedic science enlightens that all the five Natural Forces are interrelated and interact ceaselessly. In addition, The "Prithvi (Earth)" acts as base for exchange of energy in the ariel and underground parts of crops among "Akasha/ Aether (Solar radiation), The "Vayu (Air)", The "Agni (Fire)", and "Apah / Jal (Water)" for growth and development of crops. By understanding this exchange of energy, interrelationships, and interactions agriculture flourished in Vedic/Ancient times and farmers lived with dignity

17.5.5 Classical Micrometeorology on Exchange of Energy, Interrelationships, and Interactions between "Weather Elements" and "Soil"

- Turbulent transfer/ exchange of electrons, momentum, heat, and mass (including water vapor and gases like carbon dioxide), takes place between the "Soil" and the "Weather" in and above cultivated soils (near the ground)
- These are all generic groups of processes, by which a quantity; electrons (solar radiation), momentum (air), heat (temperature), matter/mass (water) , is transferred from high electric potential, high concentration, high temperature, and high fluid velocity respectively to those regions where this quantity is lower

- Micrometeorological knowledge about energy exchanges and transport near the cultivated soils have useful applications in agriculture and livestock systems. Example: With the knowledge of energy exchanges between the cultivated soil and weather near the ground it is possible to modify the moisture status, temperature of upper soil layers, etc., at micro level for successful crop production
- The transfer/exchange processes are scientifically explained by different modern physical laws in classical agrometeorology and are being successfully used in solving several field problems for crop production
- If new micrometeorological technics and technologies are developed to solve climate change issues on agriculture these will certainly prove successful for sustainable agriculture, food production, and food security under climate change

17.5.6 Crisp Review of Knowledge on Classical Micrometeorology for Climate Change Studies

- In the "Soil, plant and weather continuum (SPWC)" achieving and maintenance of soil health, plant health, and weather health is possible through micrometeorological technics, which is the most essential need of the hour under climate change scenarios
- According to both Vedic and classical agrometeorology the weather elements/natural forces are renewable energy sources and non-monetary
- Therefore, at a time when the cost of cultivation of crops and rearing animals is tremendously increasing beyond the control and affordability of farmers and climate change is further exacerbating the issue, micrometeorological technics and technologies thus developed will be gamechangers in successful agriculture, food production, and food security
- Therefore, micrometeorological technics and technologies may be developed to address the challenges of climate change for successful crop production

- Generalized vertical profiles of weather elements and carbon dioxide in crop canopy
    - Within a crop canopy, vertical profiles of weather elements and carbon dioxide concentration exhibit characteristic patterns due to the interactions between the weather and crops **(Fig:17.1)**

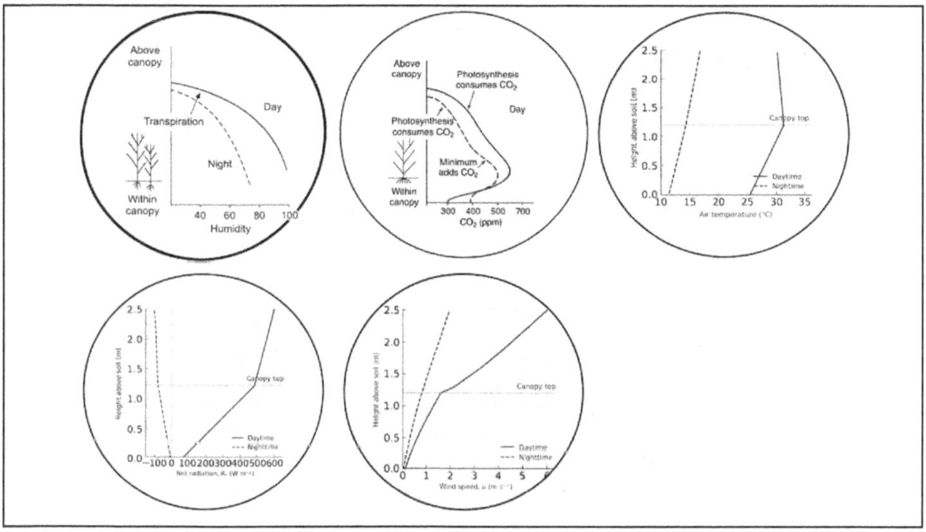

Fig. 17.1 Generalized vertical profiles of weather elements and carbon dioxide in crop canopy.

Source: Unknown. Class lecture notes

- Humidity: Relative humidity tends to be higher within the canopy compared to the ambient air due to plant transpiration and reduced wind speeds within the denser foliage
- Carbon dioxide content: Carbon dioxide content during the day tends to be lower within the canopy due to uptake by plants during photosynthesis, potentially reaching a minimum at a certain height within the canopy
- Air temperature: A daytime temperature gradient usually exists within the canopy, with higher temperatures near the top and progressively cooler conditions towards the bottom, where shading reduces solar radiation and potentially traps

cooler air. At night, the canopy can decouple from the overlying weather, with radiative cooling causing cooler temperatures within and above the canopy

- Net radiation: Net radiation is significantly reduced during the day due to interception by crop canopy
- Wind Speed: Wind speeds are significantly reduced within the canopy compared to above-canopy levels due to the drag exerted by leaves and branches
- These vertical profiles are dynamic and influenced by several factors such as
  - Solar radiation penetration: Affects temperature, humidity, and the strength of the carbon dioxide gradient by impacting photosynthesis
  - Canopy density and structure: Influences wind speed reduction, turbulence, and the penetration of light and wind
  - Crop type and stage of development: Leaf area index, architecture, and physiological processes (transpiration and photosynthesis) change with crop development impacting the microclimate and carbon dioxide profiles
  - Soil moisture availability: Influences transpiration thereby humidity levels within the canopy
  - Weather conditions: Turbulent mixing and the overlying atmospheric boundary layer conditions also play a role in shaping within-canopy profiles
- Understanding these vertical profiles is crucial for weather modification at micro level. Also, to accurately model and predict crop growth, water use, and carbon exchange between crops and the weather under climate change scenarios

- Classical micrometeorology of "Soil, plant and weather continuum (SPWC)"
  - The soil-plant-weather continuum (SPWC) describes the continuous movement of water from the soil, through the plant, and into the weather **(Fig:17.2)**

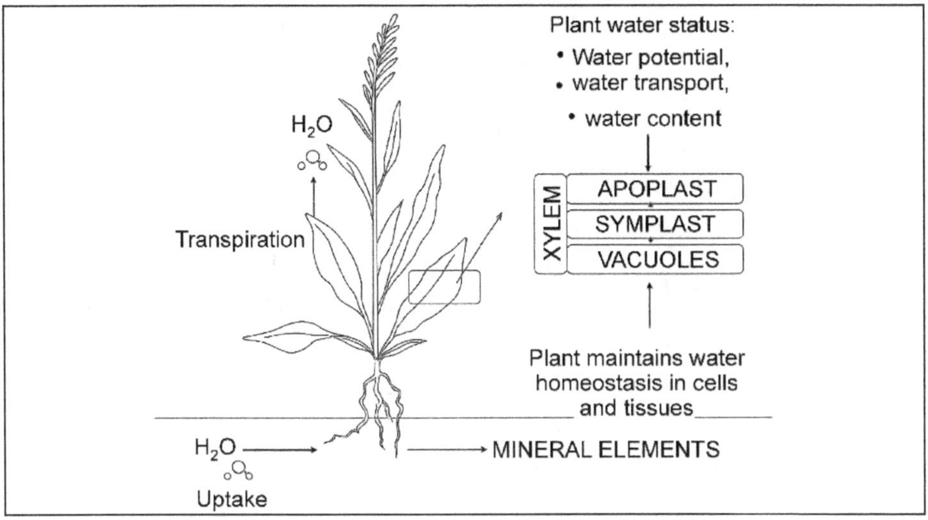

Fig. 17.2 Soil plant weather continuum.
Source: Levinsh (2023)

- o Water flows along a gradient of decreasing water potential which acts as the driving force for movement
- o Weather: A low (more negative) water potential in the weather, caused by low humidity, pulls water vapor out of the plant's leaves through transpiration
- o Plant: This transpiration creates negative pressure or tension in the xylem, the plant's water transport tissue. Due to water's cohesive properties, this tension pulls water up from the roots in a continuous column
- o Soil: The roots absorb water from the soil in response to the pressure gradient created by transpiration. Water moves from the higher (less negative) potential in the soil to the lower potential in the roots
- o Linkages and homeostasis
- o Plant water status
    - ▪ Regulation: Plants maintain water balance (homeostasis) by controlling water loss through the stomata. Guard cells regulate stomatal opening and closing, balancing the need for carbon dioxide uptake for photosynthesis with the need to conserve water

        - Strategies: Plants can be categorized as either "isohydric," which tightly regulate water potential and transpiration, or "anisohydric," which allow water potential to drop more freely
    - Soil mineral and water uptake
        - Combined uptake: As roots absorb water, they also take up dissolved mineral nutrients from the soil solution via osmosis, diffusion, and active transport. This process is known as mass flow
        - Osmotic balance: The active transport of mineral ions into root cells lowers the osmotic potential inside the root, which helps drive water inward via osmosis
        - Nutrient availability: If soil is too dry, mineral nutrients cannot dissolve or be transported effectively, hindering plant uptake despite their presence in the soil
    - Homeostasis and the continuum
        - Dynamic feedback loop: The SPWC is a dynamic system. A plant's ability to maintain water homeostasis is directly linked to the water availability in the soil and the evaporative demand of the weather
        - Stress response: Under drought stress, decreasing soil water potential forces a plant to close its stomata, reducing photosynthesis to prevent a lethal drop in plant water status. This can impact mineral uptake, growth, and survival
        - System resistance: The entire pathway from soil to weather offers hydraulic resistance to water flow. Under stress, high resistance in the soil or within the plant (due to xylem cavitation) can disrupt the continuum, threatening the plant's water balance
- Examples of classical Micrometeorological technics
    - Evapotranspiration (ET) Modeling: To estimate crop water needs for efficient irrigation, models like Penman–Monteith are widely used to calculate reference evapotranspiration (ETo), which is then adjusted with crop-specific coefficients (Kc) to estimate actual crop water use (ETc). Penman–

Monteith consistently outperforms other models across seasons, weather, and climates

- To create favorable growing conditions and mitigate weather extremes methods such as shelterbelts and windbreaks are used. These reduce wind stress and evaporation, and moderate temperature and conserve soil moisture
- Controlled Environments (greenhouses): Precisely regulate temperature, humidity, and CO_2 levels.
- Smart Sensors and IoT Integration: In real-time monitoring and decision-making, tools such as wireless sensor networks track micrometeorological variables and feed data into automated irrigation and climate control systems
- Field-Based Validation Tools such as smart field weighing lysimeters directly measure ETa (actual evapotranspiration) to validate model predictions
- Heat trapping for Microclimate modification
  - Heat trapping at a micro level using furrows and stones is most useful micrometeorological technic under changing climate **(Fig:17.3)**
  - Microclimates and thermal mass: The techniques described create localized environments known as microclimates, which can be warmer than the surrounding area. This warmth is amplified by incorporating materials with high thermal mass i.e., they can absorb, store, and release heat efficiently
  - Sunny-side planting and furrows: Planting on the sunny (south or west-facing in the Northern Hemisphere) side of a furrow maximizes exposure to solar radiation. The furrow itself can help trap and direct this heat towards the plants and soil. The crop plants on sunny side grow taller than when planted on opposite side

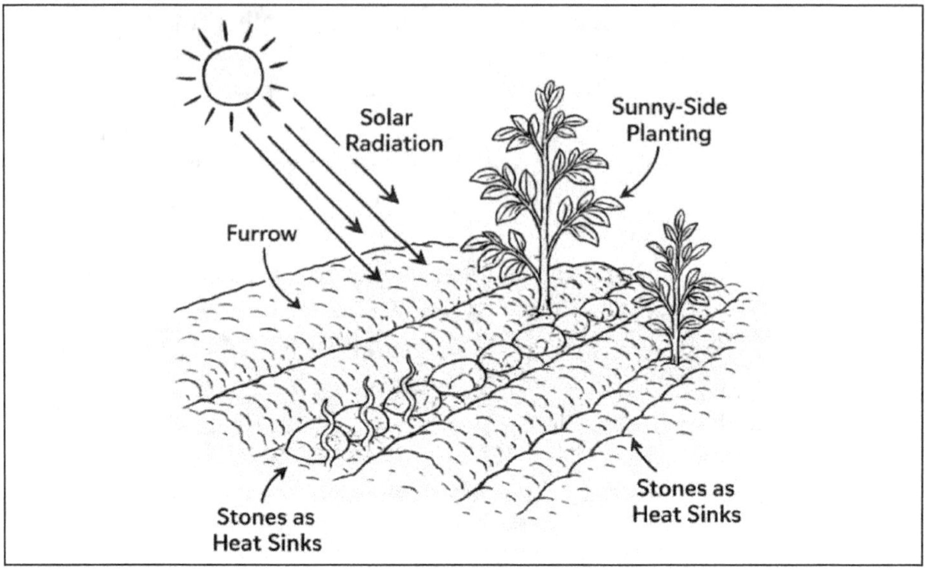

Fig. 17.3 Heat trapping for microclimate modification.

Source: Unknown. Class lecture notes

- Stones as heat sinks: Strategically placed small stones or other hardscaping materials (bricks or concrete) between plants act as heat sinks. They absorb solar energy during the day and radiate this stored heat back into the surrounding environment at night, moderating temperature fluctuations and reducing the risk of frost damage
- Benefits: These methods can extend the growing season, potentially allowing earlier planting and later harvesting. It can also create a more favorable environment for heat-loving plants that might otherwise struggle in a particular climate. This approach is particularly effective in areas with significant temperature swings between day and night

- Regulating the effect of incoming energy
  - Placing an insulation layer on or near the ground surface. Example: Paper, Straw, Polystyrene, Trees, etc.
  - Heating
  - Changing the absorptivity of ground

- Varying the air temperature by using wind machine or erection of a shelterbelt
- Example: Adjusting canopy density of grape vine plantation to regulate fruit exposure to sunlight

o Modification of the thermal properties of the ground

- By increasing the absorptivity of the ground
- By changing the thermal conductivity by cultivation, rolling, and irrigation
- By altering the heat capacity by adding or draining water
- By varying the rate of evaporation
- Examples: Removal of weeds, regulating soil moisture, placing mulch, screens, sand on the ground surface
- Sloping vineyards can offer advantages in terms of drainage and sun exposure, with southern slopes in the Northern Hemisphere often preferred for maximizing heat accumulation
- The grape plantations near large bodies of water can moderate temperature fluctuations, extending the growing season and potentially preventing vine damage
- Areas near large bodies of water like lakes, can benefit apple orchards by moderating temperatures, reducing frost damage in the spring, and preventing extreme heat in the summer
- Use of shade cloths to protect apples from sunburn during extreme heat events

o Modification of solar radiation

- Changes of surface geometry,colour, slope etc
- Crop management through changes of layout

o In summary, micrometeorological technics are helpful to improve irrigation efficiency and conserve water, enhance crop yield and quality, reduce vulnerability to pests, diseases, and climate stress, support sustainable agriculture through data-driven decisions

- **Inference:** The modern Agrometeorologists proved that turbulent transfer/ exchange of electrons, momentum, heat, matter/mass (including water vapor and gases like carbon dioxide), takes place between the "Soil" and the "Weather" in and above cultivated soils (near the ground). Micrometeorological knowledge about these energy exchanges and transport near the cultivated soils have useful applications in agriculture and livestock systems

17.5.7 Summary of Analogies and Transfer Processes between "Vedic Knowledge" and "Classical Micrometeorology Knowledge"

- Depending upon the purpose and perception, the word "Soil" has different meanings for different people. In the Oxford Dictionary, "Soil" as a noun connotes the upper layer of the earth in which crops grow. In reverence for this natural resource and to make the term convey dynamism, "Soil" is considered as "Soul of infinite life." In Indian native perception, soil is considered the "Origin of life". From the point of view of agricultural use, soil is the loose earth surface that supports crops, its quality determining the type and density of vegetation. Farmers through the ages have treated soil as a "Living Entity" and "Mother Earth"
- Soil serves six functions: (a) base for energy interactions among weather elements and produce biomass (b) performs filtering, buffering, and transformation actions among weather elements (c) a source of biodiversity (d) mechanical support to the crops and infrastructural development (roads, railways, waterways, homes, farm infrastructure etc.) (e) supplies geogenic/ pedogenic raw materials, and (f) preserve cultural heritages of landform/ landscape
- The MCCC considered The "Panch Bhutas" viz., Akasha / Aether as "Solar Radiation"; The Vayu (Air) as "Air" The Agni (Fire) as "Temperature"; "Apah / Jal (Water)" as "Rainfall, Dew, and Fog "; Prithvi (Earth) as "Cultivated Soil" **(Table. 17.1)**
- The MCCC also considered that the dynamic transfer/ exchange of energy, momentum, heat/thermal energy, and matter between the "Soil" and the "Weather" includes interaction among Akasha/Aether as "Solar Radiation"; The Vayu as "Air" The Agni (Fire) as "Temperature"; "Apah/ Jal (Water)" as "Rainfall, Dew, and Fog with Prithvi (Earth) "Cultivated Soil" as the foundation and base

Table. 17.1 Summary of analogies and transfer processes between "Vedic knowledge" and "Classical micrometeorology knowledge"

S.no	Rate process	Driving force	Conductance (transport property)	Law of molecular transfer
1	Solar energy transfer (solar radiation). In the concept "Akasha/Aether "	Photon Concentration gradient in chloroplast during photosynthesis	Radiant flux density	Energy transfer through Photons. The Wein's Law, Kirchoff's Law, Stefan Boltz's man law, etc.,
2	Momentum transfer (air). In the concept "Vayu"	Velocity gradient	Viscosity (kinematic viscosity)	Newtons' Law of viscosity
3	Heat transfer (temperature). In the concept "Agni "	Temperature gradient	Temperature gradient (thermal conductivity, thermal diffusivity, etc.)	Fourier's Law
4	Mass transfer (rainfall/ precipitation/dew). In the concept "Apah or Jal "	Concentration gradient	Mass diffusivity	Fick's Law
5	Combined transfer processes among Solar, momentum, heat, and mass, takes place in aerial and underground parts of "Cultivated soil", the soul of infinite life. In the concept "Prithvi "	Combination of all the above, individually, collectively, and simultaneously	All the above exchange/ transport processes	All the above four Laws acting collectively, individually, collectively, and simultaneously in cultivated soils

- **Inference:** Both the Vedic and classical agrometeorological sciences proved that the phenomenon of dynamic exchanges of electrons, momentum, heat, and matter/mass between the "Soil" and the "Weather" includes interaction among "Solar Radiation/Aether (radiant energy), Wind/Vayu (momentum), Temperature/Agni (thermal energy /heat), and Water/Apah (molecules as catalyst)" with "Soil" as foundation/ base. This knowledge is useful for developing micrometeorological technics and technologies to solve problems under changing climate

17.5.8 Earliest Investigations and Research on Crop Weather Relationships and Microclimate of Crops in India

- Royal Commission on Agriculture (RCA) during 1920s: The need for close collaboration between the Departments of Agriculture and Meteorology in India was felt by the Royal Commission on Agriculture during 1920s to better understand the relationship between weather and crops, so that disruption and damage caused to agriculture, economy, and human life by weather hazards like droughts etc., which could be reduced, marginalized or even controlled from turning into famine
- Earliest investigations taken up by Agricultural Meteorology Division in Pune since 1932:

  The first Director of Agricultural Meteorology, Dr. L. A. Ramdas had done extensive research and published number of research papers in Agricultural Meteorology. The investigations led by him include " Microclimate of crops"
- Research in Agrometeorology in ICAR under AICRPM at ICAR- CRIDA since 1983: The third of the four objectives was "To evaluate different techniques of modification of microclimate for improving the water use efficiency and productivity of crops"
- **Inference:** Way back in 1920s the RCA, the research led by Dr. L. A. Ramdas since 1932, and research on Agrometeorology since 1983 under AICRPM at ICAR-CRIDA recognized the importance of research on crop weather relationships and micrometeorology of crops for food production and food security

17.5.9 Process and Procedure of Decision making on "Climate Policy" at SAUs level

- The core points and main text for consideration in decision making and for inclusion in the "Climate policy" and for enacting additional Act and subsequent Statutes into the existing Act and Statutes of SAUs with reference to "Microclimate Concept for Climate Change (MCCC)" are
  - Simultaneous implementation of PCCC and MCCC along with climate change and agriculture center (CCAC) as detailed under Chapter 18 of the book helps the farmers to reduce their cost of cultivation of crops and livestock rearing by 50 per cent in addition to an increase in the crop and livestock yields by 50 per cent, and 10 per cent increase in quality of produce under climate change scenarios
  - The main point with reference to MCCC that may find place in "Climate Policy" and in additional Act and subsequent Statutes of SAUs is "Micrometeorological technics and technologies as appropriate for climate change be continuously developed and provided to all stakeholders of agriculture on priority to counter the adverse impacts of climate change"
  - The Vedic science enlightens that all the five Natural Forces are interrelated and interact ceaselessly. In addition, The "Prithvi (Earth)" acts as base for exchange of energy in the ariel and underground parts of crops among "The Akasha/Aether (Solar radiation), The "Vayu (Air)", The "Agni (Fire)", and "The Apah/ Jal (Water)" for growth and development of crops. By understanding this exchange of energy, interrelationships, and interactions the agriculture in Vedic/Ancient times was successful and farmers lived with dignity
  - The modern agrometeorologists proved that turbulent transfer/ exchange of electrons, momentum, heat, and matter/mass (including water vapor and gases like carbon dioxide) takes place between the "Soil" and the "Weather" in and above cultivated soils (near the ground). Micrometeorological knowledge about these energy exchanges and transport near the cultivated soil (micro level) have useful applications in agriculture and livestock systems

  - Both the Vedic and classical micrometeorological sciences proved that the phenomenon of dynamic exchanges of electrons, momentum, heat, and matter/mass, between the "Soil" and the "Weather" includes interaction among "Aether/Solar Radiation (radiant energy), Vayu/Wind (momentum), Agni/ Temperature (thermal energy /heat), and Apah/ Water (water molecules as catalyst)" with "Soil" as foundation/ base. This knowledge is useful for developing micrometeorological technics and technologies to solve problems under changing climate
  - Way back in 1920s the Royal Commission on Agriculture, earliest investigations taken up by Agricultural Meteorology Division in Pune since 1932 and research in Agrometeorology in ICAR under AICRPM at ICAR-CRIDA since 1983 recognized the importance of research on crop weather relationships and micrometeorology of crops
  - The MCCC is a continuation and complimentary to PCCC. These two concepts work well with one another and can make a phenomenal impact on profitable agriculture, food production, and food security under climate change
- Other points and issues for consideration in climate policy development and implementation are
  - In addition to all the points and issues mentioned with reference to PCCC for its development and implementation, the following additional points may be useful for successful development and implementation of MCCC as well
    - Development of necessary technical infrastructure for research on micrometeorological technics
    - Examples: Best practice setup (for a comprehensive study on crops): A combination of AWS (weather baseline); eddy covariance system (fluxes); soil sensors (moisture + heat flux); canopy PAR + infrared canopy thermometers; data logger with telemetry. Few modern controlled environmental facilities for research on influence of changing weather on crops

  - Recruitment of additional agrometeorologists with additional knowledge of micrometeorology and contemporary computer sciences on regular/permanent basis
  - Enabling synergy between ongoing research related to micrometeorology of "Agronomists, soil scientists, plant breeders, and scientists and experts of related disciplines" and "Agrometeorologists"
  - The scientific works of these scientists shall achieve perfect soil health, plant health, and weather health which are negatively impacted by climate change, through their new "Micrometeorological technics and technologies" on "Soil, plant, weather continuum (SPWC)" for increased crop yields, enhanced food production, and food security under climate change scenarios

- The Making of additional Act and Statutes
  - May be followed as in the case of MPCC or as appropriate
- Through the above processes and procedures, the climate policy of SAUs that include "Micromet Concept for Climate Change (MCCC)" on the lines of PCCC will be in place. An additional Act and subsequent Statutes that read- Microclimate Concept for Climate Change be incorporated into the existing Act and Statutes of the SAUs which makes provision for issuing executive orders by SAUs under the "Subject" as "Micrometeorological technics and technologies as appropriate for climate change be continuously developed and provided to all stakeholders of agriculture on priority to counter the adverse impacts of climate change"
- Take Home
- ***"Sakala Samasta Sanmangalani Bhavantu"- "May auspiciousness happen everywhere"***
  - Through PCCC and MCCC a message goes to the world that "Agrometeologists in synergy with the activities of concerned disciplines of the SAUs can confidently handle the negative impacts of climate change on agriculture"
  - Simultaneous implementation of PCCC and MCCC along with climate change and agriculture center (CCAC) as detailed under Chapter 18 of the book helps the farmers to

reduce their cost of cultivation of crops and livestock rearing by 50 percent in addition to an increase in the crop and livestock yields by 50 per cent, and 10 per cent increase in quality of produce under climate change scenarios

- o Additionally, the combined impact of PCCC and MCCC leads to double the income of farmers as well under climate change

- ***"Sarve Jana Sukhino Bhavantu"*- "May all people be happy" / "Let all people be happy"**

17.6 To make a beginning from Home

- The science of modern Physics is about "Efficiency", the output to a given input
- Classical Agrometeorology relies heavily on modern Physics, and it is about using weather elements as non-monetary inputs in agriculture and animal rearing at a time when the cost of cultivation of crops and animal raring is beyond the control and affordability of the farmers and at a time when climate change is adding fuel to fire
- Solar radiation sustains organic life on the Earth. Nothing else. Agriculture is "Capture and storage of solar energy" through Photosynthesis
- Like other weather elements solar radiation is also non-monetary and renewable input in agriculture
- The process of photosynthesis involves and integrates all weather elements
- The IPCC reported that global warming due to climate change is happening, and it is unequivocal
- Climate change is negatively impacting all weather elements and consequently photosynthesis is adversely affected by abrupt changes in weather elements
- This in turn is resulting in reduced crop yields

- Therefore, it is essential that authenticated, and science-based weather and climate information, services, and products be made available on lines of PCCC and MCCC to the farmers
- It is an official process and procedure
- While this process is going on, the farmers may take initiative and follow the available information within their reach on weather and climate and start adopting the ITKs available for microclimate management
- This beginning trigger positive and impactive results
- Also, this first step helps the farmers to make a beginning in reducing the cost of cultivation of their crops and livestock by 50 per cent, additional 50 per cent increase in yields and another 10 increase in quality of their produce and double their income under climate change.

17.7 Addressing the Issues of Climate Change in Agriculture and Allied Sciences

- To address climate change challenges individuals, families, communities, governments, organizations, etc., have the best sciences available to plan for and implement climate-smart and resilient practices for agriculture and livestock. Two critically important approaches which are available to managing climate resilience are mitigation and adaptation. These strategies are inter-related and must be pursued simultaneously as both can potentially improve resilience to the changing climate.

Chapter 18

Project on Establishment of Climate Change and Agriculture Center (CCAC)

"Vedic wisdom highlights that Dharma (righteous conduct) extends to Earth's protection, emphasizing the sacred duty to be stewards of nature. Responsible policies and law makers, guided by this principle, must allocate finances to safeguard the environment, weather, and ensure global food security. By aligning actions with this reverence, sustainable practices are cultivated, the well-being of all beings is ensured, and the sacred trust to Mother Earth is fulfilled"

18.1 Introduction to Projects on Climate Change

- Scientists propose climate change projects to funding agencies for several crucial reasons. This is because they are driven by the urgency of climate change and rely on these agencies to provide the necessary financial and collaborative support to conduct research, generate knowledge, foster innovation, and ultimately contribute to a more sustainable future
- Climate change is a complex, far-reaching problem with impacts on agriculture and allied sciences, human health, economies, ecosystems, and societies worldwide

- Research is vital to understanding the complexities of climate change, developing solutions, and mitigating its effects
- Conducting climate change research, especially large-scale or technologically advanced projects, can be expensive, requiring funding for personnel, equipment, materials, and infrastructure
- Funding agencies provide this financial support, enabling researchers to pursue ambitious and impactful projects that would otherwise be impossible
- Public research funding plays a vital role in generating knowledge and innovation, including technological advancements and sustainable economic models
- Climate change necessitates an interdisciplinary approach, requiring collaboration among experts from various fields like agricultural meteorology, agronomy, soil sciences, veterinary sciences, fishery sciences, natural sciences, social sciences, agricultural engineering, humanities, etc.
- Funding agencies often encourage and incentivize interdisciplinary research proposals, promoting collaboration and a holistic understanding of climate challenges
- Research findings are crucial for informing climate policy decisions and developing effective strategies for mitigation and adaptation
- Funding agencies contribute to this by supporting research that generates evidence-based insights relevant to policymakers and the broader public driving climate action

18.2 Project Title

- Establishment of "Climate Change and Agriculture Center (CCAC)"

18.2.1 The Genesis

- Excerpts from IPCC Reports of 2021, 2022, and 2023 on the impacts of global warming of 1.5°C above pre-industrial levels and related global greenhouse gas emission pathways, in the context of strengthening the global response to the threat of climate change, sustainable development of agriculture and livestock and efforts to eradicate poverty indicate the following

- Global warming is likely to reach 1.5°C between 2030 and 2052 if it continues to increase at the current rate
- Warming from anthropogenic emissions from the pre-industrial period to the present will persist from centuries to millennia and will continue to cause further long-term changes in the climate system, such as sea level rise with associated impacts
- Climate-related risks for natural and human systems are higher for global warming of 1.5°C than at present, but lower than at 2°C
- These risks depend on the magnitude and rate of warming, geographic location, levels of development and vulnerability, and on the choices and implementation of mitigation and adaptation options
- On projected climate change, potential impacts and associated risks, the climate models project robust differences in regional climate characteristics between present-day and global warming of 1.5°C, and between 1.5°C and 2°C. These differences include increases in (a)mean temperature in most land and ocean regions, (b) hot extremes in most inhabited regions, (c) heavy precipitation in several regions, and (d) probability of drought and precipitation deficits in some regions
- By 2100, global mean sea level rise is projected to be around 0.1 meter. This will be lower with global warming of 1.5°C compared to 2°C. Sea level will continue to rise well beyond 2100, and the magnitude and rate of this rise depend on future emission pathways. A slower rate of sea level rise enables greater opportunities for adaptation in the human and ecological systems of small islands, low-lying coastal areas and deltas
- On land, impacts on biodiversity and ecosystems, including species loss and extinction are projected to be lower at 1.5°C of global warming compared to 2°C
- Limiting global warming to 1.5°C compared to 2°C is projected to lower the impacts on terrestrial, freshwater and coastal ecosystems and to retain more of their services to humans

- Also, limiting global warming to 1.5°C compared to 2°C is projected to reduce increases in ocean temperature as well as associated increases in ocean acidity and decreases in ocean oxygen levels. Consequently, limiting global warming to 1.5°C is projected to reduce risks to marine biodiversity, fisheries, ecosystems, and their functions and services to humans, as illustrated by recent changes to Arctic Sea ice and warm-water coral reef ecosystems
- Climate-related risks to agriculture, health, livelihoods, food security, water supply, human security, and economic growth are projected to increase with global warming of 1.5°C and further with 2°C
- Most adaptation needs will be lower for global warming of 1.5°C compared to 2°C
- Climate affects many aspects of plant growth and yield and hence the effects of weather and climatic elements and their extremes will significantly alter the crop productivity and in turn, the socioeconomic conditions of many societies, both developing and developed
- These changes in weather and climate also have a direct influence on the quantity and quality of agricultural and livestock produce
- This is more pronounced in developing countries, where technology generation, innovation, and adoption are slow to counteract the adverse effects of climate change

- Therefore, the interdisciplinary nature of these issues requires a long-lasting and where possible more substantial role for a "Climate Change and Agriculture Center (CCAC)" in the efforts to promote sustainable agricultural development during the 21st century and beyond under changing climate

18.2.2 The Hypothesis

- In any nation people "Live "with food produced by farmers, the pious and sacred creators and providers
- However, the cost of cultivation of crops and animal rearing is going beyond the control and affordability of noble farmers
- They are not getting returns for their investments

- It is found that this situation is due to the negative impacts of climate on agriculture and livestock
- This is leading to resort to extreme steps
- Humanity cannot tolerate this catastrophe
- It should be stopped immediately
- Highly learned and eminent crop scientists not only in India but also across the globe including His Excellency late Dr. M. S. Swaminathan observed that there must be some serious lapses in the policies and in providing weather and climate information, services, and products for the occurrence of this catastrophe
- Therefore, they emphasized that for climate change mitigation and adaptation, impact-based solutions should be provided on priority to all stakeholders of agriculture. These include
  - Development and dissemination of weather and climate information on daily weather, services, and products including agromet advisories, climate resilient agricultural practices etc.
  - Continuing the ongoing weather and climate services and products including AASs and AABs with value addition
  - Capacity development of all stakeholders of agriculture on weather and climate under changing climate
  - Also, development of new micrometeorological and agrometeorological technics and technologies and making them available to address the negative impacts of climate change on agriculture
- Therefore, the hypothesis proposed in the project is when a CCAC is established and the farmers, scientists, teachers, students, administrators, policy makers, and all other stakeholders involved in agriculture, food production, and food security and rural development in the State are provided weather and climate information, services, and products and trained as detailed above, then
  - It is possible to reduce the cost of cultivation of crops and livestock rearing by 50 percent, increase crop yields by 50 percent, and another 10 percent in quality. In addition,

doubling the income of farmers and rural development under climate change can be achieved

- These efforts would collectively enhance the climate resilient agriculture production systems, and the farmers would produce crops profitably under climate change scenarios and integrated rural development is achieved
- This enables both the farmers and people of the State "Live" comfortably with "Dignity"

18.2.3 Key Words

- Climate Change, agriculture, agrometeorology, micrometeorological technics and technologies, development and dissemination of information on daily weather, services, and products, early warnings, agromet advisories, climate resilient agricultural practices, use of weather, climate and agrometeorological knowledge as non-monetary inputs, PCCC, MCCC, weather health, rural development, ICTs, etc.

18.2.4 Project Proposer, Implementing Institution, and Collaborating Institutions

- The project proposer: The Registrar of the SAU/As appropriate
- Implementing institution: The SAU
- Collaborating institutions
  - MOUs at Global level will be established with, World Meteorological Organization (WMO), Geneva, Switzerland; National Weather Service (NWS)- USA; European Centre for Medium Range Weather forecasts (ECMWF), Europe; UK Met Office- United Kingdom; Meteo France- France; Japan Meteorological Agency (JMA)- Japan; Bureau of Meteorology (BoM) – Australia; Environment and Climate Change Canada (ECCC)- Canada; Deutscher Wetterdienst (DWD), Germany; Global Private weather companies viz., The Weather Company (IBM), Accu weather, Tomorrow.io ; Overseas Universities, Institutions etc.

- MOUs at National and State level will be established with, India Meteorological Department (IMD), New Delhi; Indian Space Research Organization (ISRO) New Delhi; National Center for Medium Range Weather Forecasting (NCMRWF), New Delhi; Indian Institute of Tropical Meteorology (IITM), Pune; State Agricultural Universities (SAUs) and Krishi Vignan Kendras (KVKs), Research Institutions and Universities; IITs (Delhi, Pune, Bhuvaneswar) and other institutions which also develop new models, conduct climate studies, and improve forecast technology; Central Water Commission, New Delhi; Disaster Management Authorities (NDMA, SDMAs); Private weather companies viz., Skymet weather, Weather Risk Management Services (WRMS), Earth Networks India etc.

18.2.5 Budget

- Approximately Indian Rs 100.0 crores / 11.4 million US$ per annum
  - The cost of establishing CCAC at the H.Q. of the SAU, staff salaries, developing software, trainings, etc. @ Rs. 50 crores per year
  - The cost of establishing 10 sub- centers of CCAC, staff salaries, developing software, trainings etc. Rs.50 crores per year

18.2.6 State of Knowledge

- In addition to the information provided under chapters from 1 to 17 in the book and on "Introduction to Projects on Climate Change, "The genesis" and "The hypothesis" as above, a crisp and brief additional state of knowledge pertaining to the project is as follows
- Farmers are the backbone of economy of any State. The entire State and the Country depend on them for food production and food security. However, climate change is negatively impacting agriculture and farmers are not only encountering huge losses but also resorting to extreme steps. When the farmers lose

confidence in their noble profession and resort to extreme steps, surmounting this tragedy is a daunting task to tackle by all stakeholders

- Farmer suicides have been a persistent concern in India for decades, reflecting agrarian distress. According to NCRB (National Crime Records Bureau) 2022 data 11,290 farmers and cultivators died by suicide in 2022 (a rise from 10,881 in 2021). Together with agricultural laborers, the total suicides in the farming sector reached 11.2% of all suicides in India
- Government responses that are yielding desired results: Loan waivers (state-level), Crop insurance schemes (e.g., Pradhan Mantri Fasal Bima Yojana), Income support (PM-KISAN, ₹6,000 annually to farmers), Focus on irrigation, MSP hikes, Kisan Credit Cards, and rural employment (MGNREGA)
- Need for Weather and Climate information and Services in Agriculture
- Rising Climate Risks: Agriculture is highly sensitive to temperature shifts, erratic rainfall, droughts, floods, and extreme events that are intensifying with climate change
- Food Security: Reliable weather and climate information helps farmers secure yields, reduce losses, and stabilize food supply chains
- Efficient Resource Use: Forecast-based agroadvisories support water-saving irrigation, optimized fertilizer use, and timely pest/disease control
- Risk Management: Climate services underpin crop insurance, early warning systems, and disaster preparedness, reducing economic vulnerability
- Sustainability and Adaptation: Seasonal forecasts and climate projections guide crop choices, planting dates, and land management fostering long-term resilience
- Policy and Investment Support: Governments and agribusinesses rely on climate services for planning, infrastructure design, and climate-smart agriculture programs
- Therefore, providing weather and climate information and services are vital for helping farmers and policymakers adapt to climate change, manage risks, protect livelihoods, and ensure sustainable agriculture, food production, and food security

- Agrometeorological Insights
  - Rising Temperatures and Heat Stress: Global mean surface temperatures are rising faster than previously projected in many agricultural regions. The research findings on this issue are; Heat stress reduce yields of major crops especially during flowering and grain-filling stages. Increased livestock heat stress reduces milk production and fertility. Recent findings (2023–2025) of multi-model simulations show yield declines of crops under non-adaptive management. Heatwave frequency is rising in South Asia and Africa, sharply increasing risks of crop failure
  - Shifts in Rainfall and Water Availability: The research findings on this issue are; Intensified hydrological cycles, with more frequent droughts and extreme rainfall are observed; Drought stress lowers soil moisture and reduces planting window stability. Intense rainfall increases soil erosion, nutrient leaching, and flood risks. Studies in 2024 highlight growing importance of "flash droughts" (rapid-onset droughts) in the U.S. Midwest, Southern Africa, and China
  - Carbon Dioxide (CO_2) Fertilization vs. Nutrient Dilution: The impacts include gains are limited by nitrogen and phosphorus availability. Elevated CO_2 reduces grain protein and micronutrient content (iron, zinc). Recent Findings of 2023 FACE (Free-Air CO_2 Enrichment) experiments show yield increases plateauing and nutrient dilution is persistent. Breeding programs now focus on nutrient-density under high CO_2
  - Pest, Disease, and Weed Dynamics: The trends indicate that warmer, wetter climates expand the range of pests and pathogens. The impacts include longer growing seasons and multiple pest generations. Herbicide resistance spreading faster in weedy species. Recent findings of 2024 research shows climate-driven northward expansion of fall armyworm in Europe and Asia
  - Adaptation and Agroclimatic Zoning: Strategies emerging from recent research are climate-smart cropping calendars for adjusted sowing dates to mitigate heat/cold stress
  - Agroforestry and soil cover systems: These are found to buffer against drought and extreme heat

- o Digital agrometeorology: Use of AI, ML, GIS, and remote sensing for weather forecasting at micro level
- o Recent results of 2024 studies in Africa show sorghum and millet outperform maize under projected 2050 climates. Expanded use of seasonal climate forecasts improves farmer income when integrated with crop insurance
- o Emerging Tools: Machine learning in agrometeorological forecasting improves accuracy of yield predictions by ~20%. Remote sensing (Sentinel, MODIS, CubeSats) tracks soil moisture and evapotranspiration at field scales. AI-driven pest early-warning systems now operational in India, China, and East Africa
- o Climate change is already altering agroclimatic zones: Without adaptation, global yields of major cereals could decline. However, improved climate information,services, and products, agroecological practices, and crop breeding for stress tolerance offer pathways to resilience
- o Studies suggest that weather and climate services and products are vital for helping farmers and policymakers adapt to climate change, manage risks, protect livelihoods, and ensure sustainable food security

- Economic losses related to extreme weather events have risen 3-4 folds in several regions over the past 20 years globally
- A pioneering and trend setting weather-based technology was developed by the scientists of IARI to predict the incidence of mustard aphid a month in advance, through which farmers and policy makers are getting immensely benefited. It was reported in Coimbatore that wind is the most important factor for the dispersal and migration of some insects into an altogether new area. It was also observed that semi-log equations are best suited in explaining the relationship between crop yield and pest incidence. In India, the degree day concept has been widely adopted and used for all crops to relate crop growth, phonological development, yield, and pests and diseases. The excessively cloudy weather during wet season is often considered a serious limiting factor for rice production
- Indications that are emerging globally are; the yield of crops could be substantially increased (around 50%), cost of

cultivation of crops and livestock rearing could also be substantially reduced (around 50%) and quality of produce enhanced (around 10 %) when all stakeholders on climate change and agriculture are sensitized to use weather as a non-monetary input, follow agroadvisories, use weather products etc., in all agricultural operations. Towards this direction, use of weather and climate information, services, and products are essential

18.2.7 The Objectives

- With this background, the main objectives of the project are
- Objective 1: To establish an agrometeorology based standard "Climate Change and Agriculture Center (CCAC)" and a "CCAC System"
- Objective 2: To compile, edit, and **"Develop"** weather and climate information, services, and products on impacts of climate change on agriculture with global standard academic, research, and extension contents
- Objective 3: To develop a classical user friendly and interactive world wide web site on "CCAC System" which **"Houses"** information developed under objective 2 above and for **"Dissemination**" of the same
- Objective 4: Capacity building/development of all stakeholders of agriculture on use of weather and climate information, services, and products including value added AASs and AABs to combat negative impacts of climate change
- Objective 5: To conduct research, develop technics and technologies, and document ITKs on micrometeorology

18.2.8 Methodology

- Objective 1: To establish an agrometeorology based standard "Climate Change and Agriculture Center (CCAC)" and a "CCAC System"
  - o To mitigate the negative impacts of climate change on agriculture, to reduce the cost of cultivation of crops and livestock rearing by 50 percent, increase in crop yields by

50 percent, another 10 percent increase in quality of produce, thereby doubling the income of farmers, and for inclusive rural development under climate change, establishment of a "Climate Change and Agriculture Center (CCAC)" on lines of PCCC and MCCC is the need of the hour for the State

- Therefore, an agrometeorology science based "Climate Change and Agriculture Center (CCAC) "will be established in SAU as per the existing rules, procedures, and newly embedded Act and Statues on climate policy based on PCCC and MCCC
- The Head Quarters of CCAC will be located at the State Headquarters of the SAU
- The sub- centers will be established in 10 locations under the purview of the SAU viz., colleges, research stations, extension establishments, etc.
- All these centers come under the direct control of the Registrar of the SAU who may delegate administrative powers to the concerned as appropriate and as per need
- A high-end weather information generation and dissemination "CCAC System" powered by AI and ML with a robust ecosystem of technologies, tools, and processes that support both discovery and communication will be developed. It includes
    - Core AI/ML technologies for generating scientific information: Large Language Models (LLMs); Generative AI; Graph Neural Networks (GNNs); Deep Learning Models; and Multi-Agent AI Systems
    - Infrastructure for data collection and processing: Smart Sensors & IoT Devices; Data Lakes and Warehouses; and High-Performance Computing (HPC)
    - Dissemination and collaboration tools: Natural Language Processing (NLP); Recommendation Engines; Assistive Technologies; and Digital Publishing Platforms

        - Feedback and evolution mechanisms: Auto-Evaluation Metrics; Self-Play and Debate Models; and Human-in-the-Loop Systems

- Objective 2: To compile, edit, and **"Develop"** weather and climate information, services, and products on impacts of climate change on agriculture with global standard academic, research, and extension contents
    - At present, the level of knowledge of farmers is insufficient to cope with the new and aggravating problems under climate change. Hence, there is a clear need to provide information on weather and climate services and products for them. These include
    - Weather and climate information services
        - Services such as operational advisory services including AASs and AABs; weather forecasts for farmers; irrigation scheduling tools; pest and disease alerts; seasonal forecasts for crop planning; online Platforms and Mobile Apps; climate field schools, etc., will be provided. These services are increasingly vital for climate-resilient agriculture, helping farmers adapt to changing weather and climate patterns and reducing risks under climate change
    - Weather and climate information products
        - These help farmers anticipate and prepare for weather-related impacts under changing climate
        - Rainfall forecasts (daily, weekly, seasonal); temperature forecasts (minimum/maximum); heat and cold waves, frost alerts; drought outlooks and soil moisture forecasts; agroclimatic bulletins summarizing expected conditions for crops
        - Crop-specific advisory products tailored to the needs of crops and farming stages like planting date recommendations based on climate models; irrigation scheduling tools using evapotranspiration data; harvest timing advisories based on rainfall and temperature trends

  - Real-time monitoring tools to track current conditions and anomalies like weather station data dashboards (temperature, humidity, wind, rainfall); satellite-based vegetation indices (NDVI for crop health); soil temperature and moisture maps; growing degree day calculators for crop development tracking
  - Decision Support Systems like integrated platforms for planning and risk management which include crop yield prediction models using weather and soil data; climate-smart agriculture tools for long-term planning; livestock comfort indices (heat stress alerts); agrometeorological risk maps for floods, droughts, and frost

- Therefore, information on the above weather-related services and products useful under climate change scenarios will be compiled, edited, "Developed", and posted on the "CCAC System" and web site regularly after thorough scrutiny by groups of experts
- For this purpose, information available in textbooks/ journals/ research results, bibliography at international, regional, national, state, and local levels etc., and web sites and online sources of international and national research organizations will be compiled, edited, and developed to suit the requirements of all stakeholders
- Crop specialist scientists, experts in the disciplines from academic institutions, research establishments, extension organizations at local, state, national, regional, and international levels will be consulted in finalizing the contents
- Products such as books, booklets, pamphlets, brochures, etc., will be attractively printed. These will be distributed to all stakeholders and placed in e-form on web sites
- The contents will be made available both in English and local language
- To prepare and disseminate daily weather information, services, and products, early warnings, agromet advisories, climate resilient agricultural practices. etc., the ongoing

method on agromet advisories viz., AASs and AABs will be continued with value addition

- The quantitative district level forecast up to 5 days now being in use is a pathbreaking initiative in 2008 by GOI/IMD on weather and climate information, services, and products. This forecast system is being phenomenally improvised
- The weather parameters covered are seven and weekly cumulative rainfall forecast is also provided. The value addition is being done by RMC and MC. The SAUs are additionally providing information on management of crops/ animals etc.
- It is proposed to add further value by integrating Agromet advisories with CCAC system and web site on which state-of-the-art information is made available to create additional awareness among all stakeholders of the information.
- It is also proposed to involve students of all agricultural sciences and agricultural polytechnics in the whole process
- By this way, technical expertise will be extended to the developers/producers of AASs and AABs for improvement of quality, content, and relevance to local farming communities
- Assistance will be given for enhanced and quick dissemination of AASs and AABs etc., in the state, by using the facilities, contents, and information on CCAC system

- Objective 3: To develop a classical user friendly and interactive world wide web site on "CCAC System" which **"Houses"** information developed under objective 2 above and for **"Dissemination"** of the same

- The major functions of CCAC system and web site are
  - To "House" weather and climate information, services, and products as detailed in objective 2 above
  - To "Disseminate" critical and timely weather and climate information, services, and products through CCAC system
- Keeping the above points in view the objective would be achieved through
- Housing the information on CCAC system and web site
  - Development of web site on "CCAC System": User friendly and interactive web site will be designed and developed on "CCAC System". The CCAC system and web site 'Houses" data on weather and climate information, services, and products (digitized and archived historical and new and current data) including AASs, AABs etc., as detailed in objective 2
- Dissemination of information from/ through CCAC system and web site
  - It is essential to "**Disseminate"** critical and timely weather and climate information, services, and products through ICTs to farmers, students, administrators, policy makers etc., of agriculture, agri related establishments, and rural development to enhance and sustain crop productivity and to reduce the negative impacts of climate change on agriculture in the State. Therefore, to address these issues, farmer friendly and interactive agrometeorology science based web site will be developed on CCAC System
  - It is important that farmers get weather and climate information, services, and products for use under climate change scenarios in easily understandable form and in time through a quick communication and dissemination system. In addition, effective communication on climate change driven disasters, risks, and uncertainties is a major challenge in the State
  - Therefore, the dissemination of information from/ through CCAC system and web site involve synergy with local TV channels, FM radio; publications, MDWA and MSPT; other contemporary products such as handheld geo-positioning devices, bulk SMS in local and English languages; in

addition, voice SMSs for specific purposes viz. pests & disease control; handouts and pamphlets on IPM, INM etc., will be made available to all stakeholders

- o The internet, digital satellite technology, wind-up machines, computers etc., are new possibilities for rural areas in the State. The mobile phones, facsimiles, e-mail, wireless technologies, etc., which are available in the State offer greatest potential. Internet can accomplish accurate, timely, useful, and cost-effective information to the rural areas
- o More importantly, the decision support tools will be made available to farmers, students, administrators and policy makers in user friendly, timely, and easy formats in both local and English languages
- o The CCAC system and web site help to bridge the gap between the "Farming community" and "Sources of weather and climate information, services, and products ". The CCAC system and web site take advantage as well of extensive network of international, national, regional, district, and local establishments, agricultural educational institutions, farmer's organizations, and NGOs striving for farmers wellbeing to help disseminate this information

- Objective 4: Capacity building/development of all stakeholders of agriculture on use of weather and climate information, services, and products including value added AASs and AABs to combat negative impacts of climate change
  - o Capacity building/development of all stakeholders viz., farmers, teachers, scientists, extension specialists, policy makers, etc., on weather and climate related impacts of climate change on agriculture is essential for profitable and sustained agriculture under changing climate. This helps them build their capacity and make them updated on how to combat the negative impacts of climate change on agriculture and livestock
  - o Regular training will be organized in CCAC (on campus) and through roving seminars on "Weather- Climate-Farmers" (off campus) on "Weather- Climate- Farmers" to all stakeholders of farming and rural development on climate change, agrometeorology, and weather/ climate related issues

  - Valuable academic, research, and extension content on influence of weather / climate on farming; weather/climate services and products; weather and climate terminology, natural disasters, climate change and agriculture available on CCAC system and web site will be used for this purpose as per the needs of the stakeholders
  - In addition, standard capacity building programs on development and dissemination of daily weather, services, and products, early warnings, agromet advisories, climate resilient agricultural practices, etc., to all stakeholders will also be organized regularly in collaboration with experts from international, regional, national, state, and local level institutions, establishments, and organizations
- Objective 5: To conduct research, develop technics and technologies, and document ITKs on micrometeorology
  - There is a felt need to develop micrometeorological technics and technologies for the benefit of farmers in the State for climate resilient agriculture. This is because weather/climate can't be modified at meso and macro levels. However, it is very much possible at micro level
  - Therefore, micrometeorological technics and technologies which have proved successful since Vedic/ Ancient/ (ITKs on micrometeorology) in solving the problems of crops and livestock will be documented in addition to development of new technics and technologies to address the negative impacts of climate change
  - Research on micrometeorology will be conducted in synergy/collaboration with the ongoing research activities of the SAU and technics and technologies will be developed. This collaboration/synergy is essential as scientists of other disciplines also work for development of technologies on climate change from the perspective of their sciences. Therefore, research to develop micro-meteorological technics and technologies will be carried out in collaboration with ongoing research activities of the SAU

18.2.9 Expected Outcomes and Utility of the Products Developed

- Outcome and Utility 1: A completely integrated State level agrometeorological science based worldclass academic, research, extension, and training establishment i.e., "Climate Change and Agriculture Center (CCAC)" with state of the art discovery and dissemination system i.e., "CCAC System" will be in place at SAU Head Quarters and sub- centers in 10 locations. These state-of-the-art facilities meet all the requirements on weather and climate of farming community and all stakeholders of agriculture including livestock for profitable agriculture, food production, and food security under climate change
- Outcome and utility 2: Availability of high quality and science based global standard academic, research, and extension scientific content and information on weather and climate, services, and products. The information, products, and services housed on CCAC system and web site assist all the stakeholders of the crop production systems in the State for profitable, climate resilient agriculture, food security, and rural development under climate change scenarios
- Outcome and utility 3: Availability of worldclass CCAC system, web site, ICT platforms and weather-based decision support systems. These most contemporary and classical tools help to "House" and "Disseminate" all the required weather and climate information, services, and products including value added AASs and AABs to all the stakeholders to comprehensively, impactfully, and positively address negative impacts of climate change on agriculture and livestock
- Outcome and utility 4: Improved capacity and awareness of farmers and all stakeholders on weather and climate information, products, and services. This helps all stakeholders to thoroughly address, avoid, mitigate, and adapt to all negative impacts of climate change on agriculture, food production, and food security in addition to managing weather-related disasters, risks, and uncertainties at the production level itself thereby achieving rural development

- Outcome and Utility 5: The ITKs, new technics, and technologies on Micrometeorology. The ITKs, technics, and technologies developed on micrometeorology provide proved and tested scientific services and products that help to systematically address the negative impacts of climate change on agriculture and allied sectors thereby achieving inclusive rural development
- Finally, the above-mentioned outcomes and their utility ensure profitable and sustained agriculture and inclusive rural development. Strengthening the confidence of the farmers thereby not resorting to extreme steps. Improve their overall livelihoods and development under the inevitable climate change scenarios. These efforts on lines of PCCC and MCCC enhance crop and animal yields by 50%. There will be a minimum of 50% reduction on cost of cultivation of crops and livestock rearing. The quality of farm products will be enhanced by 10% under the climate change scenarios, thereby doubling the farmers' income.

References

The author used information from anonymous authors, online information provider sources, and publishers in preparing his lecture notes during his over three and half decades of teaching, research, and extension. The same was used in this book. However, the exact sources were not available. Therefore, the author expresses his deep sense of gratitude and indebtedness to all of them in letter and spirit. In addition to so many books; lecture notes; catalogues; cyclostyled papers; WMO, FAO, IPCC, UNEP, and IMD manuals; monographs; instruction bulletins; ICAR publications; research notes available in the library, the following publications have been consulted in this academic work to provide realistic solutions by operational agrometeorologists for the benefit of the global poorest of the poor farmers under climate change.

- Aggarwal, S., & Upadhyay, R. C. (2013). Impact of climate change on livestock production. Indian Journal of Animal Health, 52(1), 1-13.
- Ansel Farm. (2023). The Effects of Different Climates and Weather Conditions on Livestock. Retrieved from https://anselfarm.com/the-effects-of-different-climates-and-weather-conditions-on-livestock/ (Accessed June 30, 2024).
- Blick, T., Schüller, L. K., Staufenbiel, R., & Gauly, M. (2018). The impact of heat stress on reproductive performance of dairy cattle: A review. Tierärztliche Praxis Großtiere, 46(04), 213-221.
- Chakravarthy N.V.K. and Gautham R.D. (2002). Forewarning mustard aphid. NATP Mission mode project on "Development of weather based forewarning systems for crop pests and diseases. (F.No.27/99 NATP/MM-111-17 programme, Dept. of Agril. Physics, IARI, New Delhi
- FAO (2015) Climate change and food security: risks and responses. Food and Agriculture Organization of UN, | 2015, ISBN 978-92-5-108998-9

- GAMP (2010). Guide to Agricultural Meteorological Practices (GAMP), 2010 Edition. WMO-No. 134, Geneva, Switzerland.
- GRID-Arendal. CCCC Kick the Habit: A UN Guide to Climate Neutrality, text, 2008; Writer: Alex Kirby. UNEP/GRID-ARENDAL (Editorial team: Jasmina Bogdanovic, Claudia Heberlein, Otto Simonett, Christina Stuhlberger); Carto-graphics; Emmanuelle Bournay; Copy editing by Harry Forster, Interrelate Grenoble. (https://digital.library.unt.edu/ark:/67531/metadc28573/: accessed July 8, 2025), University of North Texas Libraries, UNT Digital Library, https://digital.library.unt.edu; ISBN: 978-92-807-2926-9
- Ievinsh, G. (2023). Water Content of Plant Tissues: So Simple That Almost Forgotten? Plants, 12(6), 1238. https://doi.org/10.3390/plants12061238
- IPCC, 2021: Summary for Policymakers. In: Climate Change 2021: The Physical Science Basis. Contribution of Working Group I to the Sixth Assessment Report of the Intergovernmental Panel on Climate Change [Masson-Delmotte, V., P. Zhai, A. Pirani, S.L. Connors, C. Péan, S. Berger, N. Caud, Y. Chen, L. Goldfarb, M.I. Gomis, M. Huang, K. Leitzell, E. Lonnoy, J.B.R. Matthews, T.K. Maycock, T. Waterfield, O. Yelekçi, R. Yu, and B. Zhou (eds.)]. In Press.
- IPCC, 2022: Summary for Policymakers [H.-O. Pörtner, D.C. Roberts, E.S. Poloczanska, K. Mintenbeck, M. Tignor, A. Alegría, M. Craig, S. Langsdorf, S. Löschke, V. Möller, A. Okem (eds.)]. In: Climate Change 2022: Impacts, Adaptation, and Vulnerability. Contribution of Working Group II to the Sixth Assessment Report of the Intergovernmental Panel on Climate Change [H.-O. Pörtner, D.C. Roberts, M. Tignor, E.S. Poloczanska, K. Mintenbeck, A. Alegría, M. Craig, S. Langsdorf, S. Löschke, V. Möller, A. Okem, B. Rama (eds.)]. Cambridge University Press, Cambridge, UK and New York, NY, USA, pp. 3-33, doi:10.1017/9781009325844.001.
- IPCC, 2022: Summary for Policymakers [P.R. Shukla, J. Skea, A. Reisinger, R. Slade, R. Fradera, M. Pathak, A. Al Khourdajie, M. Belkacemi, R. van Diemen, A. Hasija, G. Lisboa, S. Luz, J. Malley, D. McCollum, S. Some, P. Vyas, (eds.)]. In: Climate Change 2022: Mitigation of Climate Change. Contribution of Working Group III to the Sixth Assessment Report of the Intergovernmental Panel on

Climate Change [P.R. Shukla, J. Skea, R. Slade, A. Al Khourdajie, R. van Diemen, D. McCollum, M. Pathak, S. Some, P. Vyas, R. Fradera, M. Belkacemi, A. Hasija, G. Lisboa, S. Luz, J. Malley, (eds.)]. Cambridge University Press, Cambridge, UK and New York, NY, USA. doi: 10.1017/9781009157926.001

- IPCC, 2023: Climate Change 2023: Synthesis Report. Contribution of Working Groups I, II and III to the Sixth Assessment Report of the Intergovernmental Panel on Climate Change [Core Writing Team, H. Lee and J. Romero (eds.)]. IPCC, Geneva, Switzerland, pp. 35-115, doi: 10.59327/IPCC/AR6-9789291691647.
- IPCC, 2023: Annex I: Glossary [Reisinger, A., D. Cammarano, A. Fischlin, J.S. Fuglestvedt, G. Hansen, Y. Jung, C. Ludden, V. Masson-Delmotte, R. Matthews, J.B.K Mintenbeck, D.J. Orendain, A. Pirani, E. Poloczanskaa, and J. Romero (eds.)]. In: Climate Change 2023: Synthesis Report. Contribution of Working Groups I, II and III to the Sixth Assessment Report of the Intergovernmental Panel on Climate Change [Core Writing Team, H. Lee and J. Romero (eds.)]. IPCC, Geneva, Switzerland, pp. 119-130, doi: 10.59327/IPCC/AR6-9789291691647.002.
- IPCC, 2023: Annex II: Acronyms, Chemical Symbols and Scientific Units [Fischlin, A., Y. Jung, N. Leprince-Ringuet, C. Ludden., C. Péan, J. Romero (eds.)]. In: Climate Change 2023: Synthesis Report. Contribution of Working Groups I, II and III to the Sixth Assessment Report of the Intergovernmental Panel on Climate Change [Core Writing Team, H. Lee and J. Romero (eds.)]. IPCC, Geneva, Switzerland, pp. 131-133, doi: 10.59327/IPCC/AR6-9789291691647.003.
- Motha, R.P., and Murthy, V.R.K. (2006). Agrometeorological Services to Cope with Risks and Uncertainties. Chapter 25. The book on Managing Weather and Climate Risks in Agriculture, edited by M.V.K. Sivakumar and R.P. Motha. WMO-IMD International Workshop on Agrometeorological Risk Management held in New Delhi from 23–27 October,2006.

- Murthy, V. R.K. (2002) Basic Principles of Agricultural Meteorology. Text book. BS Publications, Giriraj Lane, Sultan Bazar, Hyderabad A.P. India
- Murthy V.R.K. Yakadri, M. and Prasad PVV (2006). Terminology in Agricultural Meteorology and Agronomy Text book. BS Publications, Giriraj Lane, Sultan Bazar, Hyderabad A.P. India, ISBN 81-7800-037-8
- Murthy, V.R.K. (2006). Coping strategies with agrometeorological and uncertainties for water erosion, runoff, and soil loss. Chapter 19.WMO-IMD International Workshop on agrometeorological risk management. New Delhi, 23-27 October.
- Murthy V.R.K. (2007) A Report on "Weather- Climate-Farmers", WMO, Geneva, Switzerland
- Murthy V.R.K. and Reddy T.Y. (2009). The weather and climate (The sound and power) as non monetary inputs in agriculture for under developed and developing countries. Paper and poster presented in World Climate Conference- 3, Geneva, Switzerland, August 27- Spet2 2009
- Murthy V.R.K. (2010) Roving seminars on weather climate and farmers in India and Srilanka. Paper and poster presentation in CAgM of Agricultural Meteorology of WMO, at Belo Horizonte, Brazil , 14-16 July 2010
- Murthy, V.R.K. (2010). Agricultural Meteorological weather elements and their observation. Chapter 2. Guide to Agricultural Meteorological Practices (GAMP), 2010 Edition. WMO-No. 134, Geneva, Switzerland.
- Murthy V.R.K. (2013). The potential of weather health (WH) and Daily Weather and Agriculture (DWA) concepts as insurance indices. Book on "The challenges of Index based insurance for food security in developing countries. Edited. Rene Gommes and Frankois Kayitokire. European Union's Publishing office, Italy ISBN 978-92-79-27783-2
- Murthy V.R.K. (2015). Climate change- Weather health indices as tools and services in Agriculture. The Andhra Agric.J. 62 (4); 737-745, 2015
- Murthy, V.R.K. (2016). Principles and Practices of Agricultural Disaster Management. Text book. BS Publications, Giriraj Lane, Sultan Bazar, Hyderabad A.P. India, 2016. ISBN: 978-93-83635-97-9-PP243

- Murthy, V. R. K., & Sreerekha, M. (2017). Agriculture Heritage. Text book. BS Publications, Giriraj Lane, Sultan Bazar, Hyderabad A.P. India, ISBN: 978-93-86819-45-1.
- Murthy, V.R.K., Jyotibasu, Kiran kumar, S. and Sreerekha, M. Practical Agrometeorology. (2019). Text book. BS Publications, Giriraj Lane, Sultan Bazar, Hyderabad A.P. India, ISBN 978-93-5230-154-6
- NASA GISS (National Aeronautics and Space Administration and Goddard Institute for Space Studies) & NOAA NCEI (National Oceanic and Atmospheric Administration and National Centers for Environmental Information, USA (2025): Global temperature anomalies (0C) averaged and adjusted to early industrial base line (1881-1910)
- NOAA, Scripps Institute of Oceanography, San Diego. (2025): Center for Marine Biodiversity and Conservation: A joint initiative among NOAA, Scripps Institution of Oceanography (SIO), UC San Diego, and the San Diego Supercomputer Center, USA
- Parry, M. L., Carter, T. R. 1989. The effect of climatic variations on food production. In: Climate and food security. Manila, Philippines. International Rice Institute 429-449
- Ramana Rao, B. V. (1988). Operational Agricultural Meteorology. Text book. Indian Society of Agronomy. Indian Agricultural Research Institute, Pusa, New Delhi
- Rathore., L.S., Kripan Ghosh., and Singh., K.K.(2025): Evolution of agromet advisory services in India. Mausam, 76, 1(January 2025), 231-256
- Sivakumar, M. V. K., Gommes, R., and Baier, W. (2000). Agrometeorology and sustainable agriculture. Agricultural and Forest Meteorology, 103(1–2), 11–26. https://doi.org/10.1016/S0168-1923(00)00115-5
- Sivakumar, M. V. K., and Brunini, O. (2005). Impacts of present and future climate variability on agriculture and forestry in the arid and semi-arid tropics. Climate Change, 70, 31–72.
- Sivakumar, M. V. K., Motha, R. P., and Das, H. P. (Eds.). (2011). Climate change and food security in South Asia. Springer. https://doi.org/10.1007/978-3-642-16397-5

- Srinivasa Rao, Ch., Pathak, H. and Dwivedi, B.S. (2024). Agriculture Sector Contributions in Achieving Net Zero Emission by 2070. Policy paper No.13, ICAR-National Academy of Agricultural Research Management, Hyderabas,500030, pp30
- Stigter, C.J. and Murthy, V.R.K. (2004). Operational agrometeorological services for extension needs and supportive role of agricultural research. In: Strengthening Operational Agrometeorological Services at the National Level, Proceedings of the Inter-Regional Workshop, March 22–26, 2004, Manila, Philippines. Edited by Raymond P. Motha, M.V.K. Sivakumar, and Michele Bernardi. Washington, D.C.: United States Department of Agriculture; Geneva: World Meteorological Organization; Rome: Food and Agriculture Organization of the United Nations. WMO/TD No. 1277, Technical Bulletin WAOB-2006-1.
- Venkataraman, S. (2014). Principles and Practice of Agricultural Meteorology. Hyderabad: BS Publications. ISBN: 9789383635347.

Websites

- Food and Agriculture Organization (FAO): fao.org
- India Meteorological Department (IMD): mausam.imd.gov.in
- Intergovernmental Panel on Climate Change (IPCC): ipcc.ch
- United Nations Environment Programme (UNEP): www.unep.org
- World Meteorological Organization (WMO): wmo.int
- www.livingfield.co.uk

www.ingramcontent.com/pod-product-compliance
Lightning Source LLC
Chambersburg PA
CBHW071520180726
48282CB00058B/597
* 9 7 8 9 3 4 7 3 1 1 4 8 2 *